CALCULUS
AND THE COMPUTER

CALCULUS AND THE COMPUTER

SHELDON P. GORDON

SUFFOLK COUNTY
COMMUNITY COLLEGE

PRINDLE, WEBER & SCHMIDT

BOSTON

PWS PUBLISHERS

Prindle, Weber & Schmidt • 🎣 • Duxbury Press • ♠ • PWS Engineering • 🔺 • Breton Publishers • ⚙
20 Park Plaza • Boston, Massachusetts 02116

PWS Publishers is a division of Wadsworth, Inc.

Library of Congress Cataloging-in-Publication Data

Gordon, Sheldon P.
  Calculus and the computer.

  Bibliography: p.
  Includes index.
  1. Calculus—Data processing. I. Title.
QA303.G67 1986   515   85-31082
ISBN 0-87150-950-4

ISBN 0-87150-950-4

Printed in the United States of America

86 87 88 89 90—10 9 8 7 6 5 4 3 2 1

EDITOR: Dave Pallai
PRODUCTION COORDINATOR AND DESIGNER: S. London
PRODUCTION: Technical Texts, Inc.
COVER PHOTOGRAPH: Greg Bowl, Bowl/Marsel Photographers
TYPESETTING: Beacon Graphics Corporation
COVER PRINTING: New England Book Components
PRINTING AND BINDING: R. R. Donnelley & Sons Company

This book is dedicated to Craig and Kenny in partial repayment
for some of the time and attention they did not receive.

PREFACE

Calculus is widely considered to be one of the greatest intellectual achievements of Western Civilization. Its methods provide some of the most powerful mathematical tools available for solving fundamental problems in most areas of human endeavor. At the same time, calculus possesses a particularly natural cohesiveness, making it a beautifully unified subject area that provides an exceptionally solid foundation for most later courses in mathematics. In view of these achievements, a natural question that must be addressed is: Why introduce the computer into calculus and thereby change, even if minimally, the focus of the subject? There are several answers to this question, and all provided the motivation for this book.

The computer can play three significant roles in the calculus sequence:

1.   It can provide the student with enhanced insight and understanding of many calculus concepts and techniques. The student is able to find out "What happens when I use this function?" or "What happens when I change this set of parameters?" by simply pushing several keys. The computer is thus able to add an exciting new dimension to the teaching and learning of calculus.

2.   It can serve as an invaluable tool for handling practical, real-world problems that are far more complicated than the artificial situations usually dealt with in the classroom. Students of calculus should learn to turn to the computer as a tool as readily as they turn to a calculator for help with simple numerical calculations.

3.   It can provide the context in which to introduce a variety of new and useful topics into calculus. For example, the following topics of a numerical nature, which are included in this text, are especially adaptable to the computer:

- various methods for finding roots of equations,
- Newton's method for solving systems of nonlinear equations,
- construction of interpolating and other approximating polynomials,
- the trapezoid rule for evaluating double integrals,
- numerical solutions of differential equations.

This book covers all of the calculus topics that are amenable to the use of the computer. Not all of these topics are intended for coverage within a standard course. After all, the computer should enhance the course; it should not displace any significant portion of the calculus. Institutions that require a computer lab with calculus certainly could treat most of the topics, however. All instructors who want to use the computer in calculus can pick and choose the topics that are best suited to their course objectives. Some might use the computer to increase the understanding of calculus topics and concepts; others might use it to emphasize its use as a practical tool for handling difficult problems; while still others may want to introduce the computer to their students via special project assignments. Accordingly, the broad selection of topics included here is intended to provide a great deal of flexibility, for individual instructors and students.

In order to achieve maximum transferability among computer models, all programs are given in a relatively simple version of BASIC. More elegant or sophisticated versions would tend to be geared to special features of particular computer systems and, as such, would not be universally usable. However, since computer graphics have such marvelous applications in calculus, Chapter 10 is devoted to developing some notions of graphing based on the graphics system of the Apple II computer. Suggestions for converting to other computer models are included.

The book is structured to accommodate a wide range of student experience with computers:

● A preliminary chapter on the rudiments of BASIC is included for students who have little or no prior programming experience.

● In the early sections, the programming is kept to a bare minimum so as not to overwhelm inexperienced students with too much BASIC initially, especially since they are simultaneously just becoming acquainted with calculus.

● In subsequent sections, as the students' experiences grow, the length and sophistication of the programs are increased accordingly.

● Provision is made in the exercises to challenge students already proficient in programming.

The book is also designed to satisfy differing philosophies regarding the extent of interaction desired between student and computer. One view shared by many instructors is that if calculus students have not already been exposed to the computer, they will certainly need it sometime in their careers. Therefore, they should have considerable interaction with the machine. However, other instructors prefer the minimum of interaction; they want students to press the fewest possible keys necessary to have the computer produce an answer.

Accordingly, most of the programs in this book are available in two versions:

● The text itself contains "no frills" versions of most of the programs. These versions are complete and need only to be keyed into the computer. Most do not exceed 10 to 20 instruction lines so as not to demand too much student time. Yet students will have a considerable amount of interaction with the computer and will be better able to appreciate the computational procedures used by the computer. After all, the instructions (and therefore the logic and the mathematics) are in front of the student rather than hidden inside a "black box."

Each of these "no frills" programs is accompanied by a series of computer exercises designed to draw on and increase the students' knowledge of programming. These exercises contain detailed instructions and suggestions on how the program can be improved. Thus, students who already know programming will have the added advantage, incentive, and challenge to apply their knowledge to flesh out the programs by supplementing them with subtleties that are left out of the "no frills" versions.

● For instructors who are only looking for student use of software and are not concerned with programming, "canned" versions of the programs are available on adoption of the text. These fairly sophisticated versions are available on diskette. The user is prompted to supply a fairly minimal amount of input (say, to define a function and supply the endpoints of an interval), and the desired results are displayed. The "canned versions" of the programs are thus designed to meet the requirement of limited student interaction with the computer.

This book is intended to be used independently of any particular calculus textbook. However, the book is linked indirectly to both Swokowski's *Calculus with Analytic Geometry* and Zill's *Calculus with Analytic Geometry* through the problems sets and the examples. In particular, many of the same problems from these two texts are included in the problem sets to show how these specific cases can be handled by the computer.

Acknowledgements I would like to express my sincere indebtedness to a number of people whose suggestions and efforts served to improve the book significantly. In particular, I want to single out Ray Greenwell (Hofstra University), Roy Myers (Pennsylvania State University), J. Bryan Sperry (Pittsburg State University), David Hill (Temple University), and John Gregory (Southern Illinois University). Special thanks are due to my sponsoring editor, Dave Pallai, for his constant assistance and support and to the production staff, particularly Susan London and Sylvia Dovner, for the professional job they did in turning the manuscript into a book. Above all, I want to acknowledge my wife Florence Gordon (New York Institute of Technology), who not only read the manuscript repeatedly, but also had to live with my constant demands for attention, assistance, and encouragement.

Sheldon P. Gordon

TABLE OF
CONTENTS

FOUR
THE INTEGRAL/80

FIVE
THE TRANSCENDENTAL FUNCTIONS/104

CALCULUS
AND THE COMPUTER

ONE
INTRODUCTION TO COMPUTERS AND BASIC PROGRAMMING

1.1
COMPUTERS AND CALCULUS

For all of its phenomenal capabilities, a computer is nothing more than a machine that can do only three simple things: it can add two numbers, it can store numbers, and it can compare two numbers to see which is larger. Everything else it does is built out of these three operations. For example, multiplication is performed as repeated addition. Division is done using a method that involves just addition and multiplication. Even when a computer is used for operations with words, as in data processing applications, these three operations are all that are involved. Thus, a list of names is alphabetized by first converting each individual letter into a numerical code and then the respective codes are compared to see which is larger.

Each of these processes may be a fairly lengthy procedure. However, one major advantage of using a computer is its speed. It can perform millions of operations per second. The other advantage is reliability—it will consistently produce the same result for the same values and will do it correctly according to the instructions given by the computer user.

The key to using the computer, therefore, lies in the instructions we give it. To make computers usable to a wide audience, manufacturers have developed, and built into their computers, systems that enable the user to indicate his or her desires by using a small set of commands. These involve a series of programming language statements (primarily English words), a series of rules for using the commands, and a series of system commands for telling the computer how and what to do to a program. These commands are then translated automatically by the computer into the specific machine-language instructions needed to implement the desired operations. The results are then automatically translated back into words and numbers. We shall not concern ourselves with how this is actually accomplished. The interested reader should refer to a book on computer science.

Our concern here is how the computer can be used in the study of calculus. There are several different ways a computer can be profitably employed in a mathematics course such as calculus. Probably the single most obvious use is as a computational tool to solve problems. As you will see, many of the types of problems that arise in calculus involve numerical solutions and the computer can make the work much easier. This is especially important if you deal with real-world problems that are far more complicated than the relatively artificial classroom situations common to calculus textbooks.

However, there are other ways that the computer can help in calculus. For one, it frees the user from doing routine calculations and manipulations so that the user can sit back and look at the answers the computer produces. This provides the opportunity to notice patterns and interrelationships that otherwise might pass unseen in the midst of doing the work. Therefore, you can begin to see the "why" of calculus instead of just the "what." Furthermore, many of the concepts and theorems of calculus do not always agree with your intuition of what should be true. As a result, most students end up accepting such results intellectually, but do not always feel those results are right. The computer is able to perform certain calculations when dealing with these topics that will provide the emotional conviction and understanding that most students need but don't get in the normal course. In summary, the computer can provide an entirely new dimension to your understanding and mastery of calculus.

1.2
SYSTEM COMMANDS

To make use of the computer, we must learn how to communicate with it. We will use BASIC, one of the most common and easiest to learn computer languages. The commands in BASIC, together with the simple rules for applying them, will allow us to write *programs* — sets of instructions to the computer that tell it what we want it to do. We will get into writing programs later in Section 1.3. Now, though, we will see how to get the computer to handle a program once it is written. These system commands are also extremely useful in the process of writing a program and it makes sense to see them first.

We communicate with the computer via a keyboard whose keys are arranged in the same pattern as on a typewriter. When one of the keys is pressed, the corresponding character appears on the screen or the paper, depending on the type of computer being used.

ENTER or RETURN: The computer does not process what is being typed and what appears on the screen until it is entered into the computer's memory. This is done by pressing the ENTER or RETURN key — or a special symbol key, depending on the computer — at the right of the keyboard. *Every* line and command that is typed must be entered before the computer will react to it.

Suppose that we have a program already written. There are a variety of things that we might want the computer to do to or with this program.

These are indicated by a series of *system commands* that are not part of the actual program.

RUN: To have the computer execute the program, we must type the command RUN and then press the ENTER key. Without the RUN command, the computer would sit forever and do nothing with the program.

SAVE: If the program we have is worthwhile, we need a way of storing it so it can be used again in the future. This is accomplished by the SAVE command. (The precise details for using SAVE depend on the particular computer you are using—you may have an account on a large time-sharing system or you may be working with either cassette tapes or disks on a microcomputer—just check with your instructor.) Usually, the program needs to be named, and the SAVE command will look something like

```
SAVE NAME      or      SAVE "NAME"
```

The use of quotes depends on the model you are using. This command can also be used to SAVE a program that is still incomplete so you can call it back at a later date to finish it. On most computers, if you SAVE a program having the same name as a previous program, the computer will wipe out the earlier version and replace it with the current one.

OLD, LOAD, or GET: Once a program has been SAVEd for future use, we must be able to recall it. On a large system, the command to do this may be OLD, LOAD, or GET followed by the name of the program, say

```
OLD CALCULUS      or      LOAD CALCULUS
```

On most microcomputers, the corresponding command is usually the word LOAD or some variation of it followed by the program name, say

```
LOAD CALCULUS      or      LOAD "CALCULUS"
```

When an existing program has been LOADed, you probably will want to simply RUN it, and this is accomplished in the obvious way.

LIST: There are times when you may be interested in seeing the actual BASIC instructions that form the program, especially if you are still working on it. The LIST command causes the entire program to be printed out for you (on the screen if you are using a monitor or screen; on paper if you are using a teletype terminal). If you want to see only a portion of the program, that also can be arranged. One of the features of BASIC is that every line of the program must be numbered. Therefore, if you type LIST 50, then only line number 50 of the program will be displayed. If you type LIST 100–200, then all lines between 100 and 200 will appear.

Note: If you are using a microcomputer with a printer connected, there is a command that allows the output to be directed to the printer either instead of or in addition to the screen. This provides you with a means of obtaining a *hardcopy* of the program instructions, for greater ease in finding errors and for keeping copies of your work to examine when you do not have ready access to a computer. The command needed depends on the particular computer. On some models, you can simply use LLIST instead of LIST and the program or particular sections of it are then printed on paper. On the Apple, you probably use the command PR#1 to open a channel to the printer and then use LIST.

NEW: This command erases an existing program in the computer's current memory (it does not erase the stored version on either tape or disk) and allows the user to start fresh. This is equivalent to turning off the power on a microcomputer and turning it back on.

DEL or DELETE: This command followed by line numbers, say,

```
DEL 100-250      or      DEL 100, 250      or
DELETE 100-250
```

erases all the lines from 100 to 250 in the program. It does not affect versions stored on tape or disk. The form of this command varies from computer to computer, so you will need to experiment to see which one works with yours.

REPLACE: This command is available only on large time-sharing systems or disk systems. It allows the user to have a previously SAVEd version of a program discarded in favor of a newer updated version under the same program name. SAVE is used initially, and all subsequent program modifications are kept using REPLACE.

BREAK, ESC, ESCAPE, CTRL/C, or other key: Every computer provides a way of stopping a program while it is RUNning; in essence, this is a "panic button" that stops the computer in its tracks if a program is running unnecessarily or is caught in an apparently endless loop.

PAUSE, CTRL/S, or other key: The system usually provides a key or set of keys that will either temporarily stop or slow the speed of the printing on the screen to allow the user to read what is coming out. Similarly, there are commands to start the printing again, but these vary greatly from one computer to another.

DIRECTORY, DIR, CATALOG, or CAT: This command, depending on the computer system, allows you to see a list of all the programs previously SAVEd in your account or on your disk.

There are many other system commands available on most computer models, but the ones listed above are the commands we will need in this course.

Making a mistake is one of the things most people fear when they are new to computers. First of all, you should realize that almost no one writes a perfect program on the first try. There are almost always errors in the program and it simply will not function. Admittedly, experienced programmers will make errors at a more sophisticated level than a novice programmer will, but we all make mistakes. The process of correcting the mistakes is known as *debugging*.

You should also realize that a mistake in a program will not harm the computer. It may cause tremendous fits of frustration for the programmer, however, as the computer repeatedly tells you that you are making errors. The trick is that you must be extremely careful and try to maintain your peace of mind. If the computer informs you that there is an error at a given line, then you should immediately LIST that line and examine it letter by letter for some kind of typing mistake. A way to change such an incorrect line will be discussed in the next section.

Often, though, errors are noticed before they are ENTERed and these can be corrected very simply. Every computer keyboard contains a DELete or BACKSPACE (←) or RUBOUT key. Each time you press this key, it will strike out the last character (including spaces) and move the cursor back one space. Eventually, you will BACKSPACE past the error, and then you can retype the rest of the line correctly and ENTER it.

On the other hand, if you do make an error and ENTER it, several things can happen. Usually, if you do not notice the mistake and eventually RUN the program, the computer will probably not recognize what you have typed as one of the few commands it knows, and so you will have violated one of the rules of the language. The result will be a message informing you that you have made a *syntax error*. Occasionally, your error may be a recognizable command and the computer will happily respond to it—you will just get something that may be unexpected or totally wrong and what is even worse, you may not even be aware of it. For instance, you might have commanded the computer to multiply two numbers instead of adding them. The best thing you can do is think about the answers you get—just because they were produced by a computer does not make them right! You may have made a mistake in your instructions or in the data you supplied.

1.3
INTRODUCTION TO THE BASIC LANGUAGE

In this section, you will begin to learn the BASIC language. Actually, you only need a relatively small, core portion of BASIC for this book, so we omit many other topics and features.

VARIABLES Most computer languages, including BASIC, have their roots in scientific and engineering work, so their structure is essentially algebraic. Therefore, the fundamental quantity we deal with is called a *variable*. In standard BASIC (as compared to some of the special features available on individual machines), a variable is denoted by a single letter (say A, B, or X) or a letter followed by a single digit (say T3 or W9, but not P15). Other versions of BASIC allow from two up to as many as forty letters in a variable name, but we avoid this here. The computer interprets each variable name as a reference to a location or address in its memory.

One special feature of the BASIC language is that each time you RUN a BASIC program all variables are automatically set equal to zero. Therefore, if you want to work with a variable that should start with a non-zero value, you have to assign the appropriate value.

ARITHMETIC OPERATIONS The standard *arithmetic operations* in BASIC are denoted as follows:

- + for addition
- - for subtraction
- * for multiplication (5 * 3, 8 * X)
- / for division (1/2, 4/A)
- ^ or ↑ or ** for exponentiation (X^5, Y↑(-1/2))

ORDER OF OPERATIONS One of the problems some people have in performing complicated arithmetic or algebraic operations is that they apparently do the work correctly, but do not get the same answer consistently when doing a given problem. For example, in

$$(14 + 6)/2 \cdot 5 - 3 \cdot (5 - 2)$$

possible answers are $-7, 60$, and 15; however, the correct answer is 41. The discrepancies come from changing the order of operations. As we said before, one advantage of using a computer is that it will always produce the same answer. This implies that it performs the arithmetic operations according to a predetermined set of rules. In particular, a computer follows a hierarchy of operations in which all powers are evaluated first, then all products and quotients, and finally all sums and differences. Essentially, the computer processes each line of an expression three separate times, once at each level of operation. The only way to avoid this unthinking adherence to the rules is by using parentheses liberally, as was done in the example above. The computer will first evaluate any quantity inside parentheses according to the same rules of order. To get a feel for this, consider the following expression for the quadratic formula written in BASIC:

```
X = (-B + (B^2 - 4 * A * C)^.5)/(2 * A)
```

The computer will first evaluate the term in the inner parentheses, B^2 − 4 * A * C, in three stages (first the power, then the product, and finally the difference). It will then raise this quantity to the power .5 and then add −*b* to it. Finally, it will divide the result by 2*a*. If there were no parentheses around the term 2 * A, incidentally, the numerator would be divided only by 2 and the result would then be multiplied by *a*. Similarly, if the entire numerator were not enclosed in parentheses, the computer would divide 2*a* into the square root term only and then subtract *b*. These rules for the order of operations are essentially the same as those normally used throughout mathematics.

While much of a programming language such as BASIC is algebraic, there are some subtle differences. Suppose a variable x is given the value 5; we could type X = 5. Thereafter, whenever x is referred to, it has the value 5. However, we can change the value for a variable in a number of ways. At a later stage in a program, we can set X = 6. The effect of this is that the computer replaces the number 5 stored in the memory location denoted by X and puts 6 in its place. From that point on, whenever X appears, it has the value 6.

The value of x can be changed in other ways. For example, it can change through a formula. If X = A + 7 * B + 8, then its value depends on the values of A and B. Further, the following formula, which makes absolutely no sense in algebra, is perfectly legitimate and extremely useful in a programming language:

```
X = X + 1
```

Thus, if x starts with value 5, this command tells the computer to add 1 to the current value of x, making 6, and then to assign this new value to x. Thus, after this command, x will be 6 instead of 5.

LINE NUMBERS Another feature of the BASIC language is that *each line* or statement in a program *must* be numbered sequentially. The line numbers must appear at the start of each line before any of the BASIC statements. (System commands such as RUN and LIST are not parts of the program and so are not numbered.) The computer will automatically process the program commands in numerical order unless directed otherwise.

When numbering the lines in a program, it is good practice to use multiples of 10 for the line numbers. This provides room to insert additional commands at a later stage if needed. For instance, if you find that you need three additional commands between lines 70 and 80, they can be inserted as lines 72, 75, and 77. (The computer automatically inserts these extra lines in numerical order even if they are typed in following line 770.) On the other hand, if the original commands were numbered 1, 2, 3, . . . , then there would be no room between lines 7 and 8 to insert three extra lines, and you would be forced to retype almost the entire program unless your version of BASIC has a renumbering utility.

Another advantage of the numbered lines in a BASIC program is that it is easier to make changes or correct errors. For example, suppose there is an error at line 200—either a typing error or an algebraic mistake. Or perhaps you decide to change something on that line. If you simply retype line 200 completely and press the ENTER key, the computer automatically will replace the old form of line 200 with the new version. (It will not change line 200 on a stored version of the program on tape or disk—the change will have to be SAVEd again or REPLACEd.)

THE BASIC **1.4** LANGUAGE

We now turn to an introduction of the core statements of the BASIC language. Before beginning, several comments are in order. The best (if not the only) way to get a feel for the features of any computer language is by actually trying it out. If you have access to a computer, sit down and try each of the statements presented below. Be willing to experiment—find out what happens when you use the statements and what happens when you misuse them. Remember, nothing you do will harm the machine, and you will gain considerably from trying these things. You will not learn them by just reading.

The BASIC statements discussed in this section are all you will need for the computer applications throughout the rest of this book. Wherever appropriate, each topic in calculus will be accompanied by a short BASIC program that will apply the computer's capabilities to solving or demonstrating that topic. The minimum expected of you is that you type each of the programs into your computer and RUN it with different selections of functions, intervals, and points.

The programs given are actually only skeletal versions of programs that function without any frills. If you examine them, you should see clearly the methods and logic by which computer capabilities can be implemented for calculus. In addition to the programs listed throughout the book, numerous exercises suggest various enhancements and alterations to the programs. You should consider these exercises ways to develop and improve your programming skills, as well as ways to achieve a deeper understanding of some of the topics of both calculus and computing.

We now consider the actual programming commands or statements in BASIC.

LET The LET or assignment statement has two primary functions. It is used to assign values to variables, as in

```
10   LET R = 5
20   LET P1 = 3.14159
```

It also provides formulas to the computer, which is tantamount to commanding the computer to perform the associated operation, as in

```
30   LET C = 2 * P1 * R
40   LET W = (C + 3 * R)^2
```

The result of these last two statements is that the computer will (after you eventually type RUN) calculate the values

```
C = 31.4159       W = 2154.4358
```

and store them. If you type in this four-line program and RUN it, you will see that no answers appear on the screen. The LET statement only causes the work to be done—it does not cause output from the computer.

Incidentally, most versions of BASIC allow the word LET to be omitted in the command, so that we can simply type

```
30   C = 2 * P1 * R
```

PRINT The PRINT statement is used to have results printed out on a screen or paper. There are a variety of options available with PRINT to perform many tasks.

1. If you type

```
50   PRINT C
```

after the four lines from above and then RUN, the computer will respond with the value 31.4159. You can also use

```
50   PRINT C,W
```

and the computer will print the values of the two variables on the same line. Similarly, you could use

```
50   PRINT R,C,W
```

and so forth for any number of variables.

2. Another option available with PRINT is

```
50   PRINT "      [any message]      "
```

In response to this, the computer will print out the message or *string* within the quotation marks verbatim—it does not analyze the characters inside the quotation marks.

3. The two options above can be combined to have the computer print a message along with the value for one or more variables. For example, you could type

```
50   PRINT "THE CIRCUMFERENCE IS ",C
```

If you RUN the above sample program, you will probably notice that when it prints, the spacing may not be especially nice. The computer display is subdivided into a series of columns or zones (anywhere from two up to six or eight zones across, depending on the model). The use of the comma between variables or message and variables is interpreted by the computer to mean that each item is to appear in the next available zone. Thus, the output is spread across the screen. This is fine for columns of figures, but not so esthetically pleasing for sentence structure. However, if the different items are separated by semicolons instead of commas, the computer will push the output together, as is done for

```
50   PRINT "THE CIRCUMFERENCE IS ";C
```

Once we have this capability for the form of the output, we can further improve it by making it still more informative, as in

```
50   PRINT "THE CIRCLE OF RADIUS ";R;" HAS
          CIRCUMFERENCE ";C
```

Notice that R and C are both outside the quotation marks—this means that their numerical values will be printed, while the two messages or strings contained within the quotes will be printed verbatim. The resulting output would then read

```
THE CIRCLE OF RADIUS 5 HAS CIRCUMFERENCE
31.4159
```

On the other hand, if we added

```
60   PRINT "THE VALUE OF W IS ";W
```

then the W inside the quotes is part of the message and will be printed as a W; the W outside the quotes is a variable and its numerical value will be printed. We would therefore obtain

```
THE VALUE OF W IS 2154.4358
```

4. Another use of the PRINT statement is in the form

```
55   PRINT
```

Essentially, this tells the computer to print nothing and a line of nothing looks just like a blank line. Also, this line was numbered 55 so that it fits between lines 50 and 60. The effect is to skip a line between the two lines of output for a much neater, clearer appearance.

5. As a shortcut, many microcomputers accept the ? symbol as a substitute for the word PRINT. Thus, you could use

```
60   ? "THE VALUE OF W IS ";W
```

6. When using a microcomputer connected to a printer, you often want your output directed to the printer to produce a hardcopy of the results. As we saw with the LIST command, the way to engage the printer depends on the computer. On some models, you simply use LPRINT instead of PRINT with all of the above options. On the Apple, you have to include the statement PR#1 as a numbered line within the program, and all subsequent PRINT statements are then directed to the printer instead of just to the screen.

INPUT

1. While the LET statement allows you to give values to the variables in your program, it is not always the most effective way of doing this. If nothing else, it is static because the values have to be given in advance and cannot be changed without changing the program. An alternative is to use the INPUT statement in the form

```
[line #]     INPUT     [variable]
```

For example, you might use

```
10   INPUT R
```

When the computer encounters this statement, it will stop, flash a question mark or other symbol, and wait for you to type in a response — a number — and press ENTER. For instance, if you type the number 5, the computer will assign to R the value 5, and it will then (and not before then) proceed to the next line of the program.

2. A useful alternative to using the INPUT statement is to combine it with a printed message or prompt. For instance, you could use

```
10   INPUT "WHAT IS THE RADIUS? ";R
```

Notice that there is a semicolon between the quotes and the variable — it is required here. Also, if your computer flashes a question mark while waiting for a response, you do not need to include one as part of the prompt.

3. You can also use INPUT for several variables simultaneously in the form

```
100   INPUT X,Y,Z
```

The variables must be separated by commas. The computer will wait until three numbers, also separated by commas, are ENTERed and then will assign the first value to X, the second to Y, and the third to Z.

END The last line in any program should be END. It signals the computer that the program is completed:

```
999   END
```

GO TO We said before that the computer processes the lines in a BASIC program in numerical order unless specifically ordered not to. One way to alter the order is with the GO TO statement that takes the form

```
[line #]    GO TO    [line #]
```

For example,

```
200   GO TO 50
```

or

```
120   GO TO 1070
```

When the computer encounters this statement, it will go directly to the indicated line and continue the program from that line on in order. Suppose you add the line

```
70   GO TO 10
```

to the program we have been constructing. We then obtain the program shown in Example 1.1.

EXAMPLE 1.1

```
10   INPUT "WHAT IS THE RADIUS? ";R
20   LET P1 = 3.14159
30   LET C = 2 * P1 * R
40   LET W = (C + 3 * R)^2
50   PRINT "THE CIRCLE OF RADIUS ";R;" HAS
     CIRCUMFERENCE ";C
55   PRINT
60   PRINT "THE VALUE OF W IS ";W
70   GO TO 10
999  END
```

When the RUN command is entered, the computer will first ask for a value for R, perform the calculations, and output the information indicated in the PRINT statements. It then reaches line 70 and is sent back to line 10, where it will again ask for a value of R. In other words, it will continually come back to line 10, ask for and get a value for R, perform all intermediate steps, and be sent back again. This process will continue indefinitely and is known as an *infinite loop*. The problem with having such a situation in a program is that it will never stop unless you press the BREAK or ESCAPE key. A better program would provide an exit from the loop as part of the program itself. We will see how this is done below.

Before going on, let's make some minor changes in this program solely to improve its appearance. To separate the information printed on the screen during each loop, add the following lines to the program:

```
64   PRINT
65   PRINT "********************************"
66   PRINT
```

When you RUN this version of the program, you will see a striking difference in the effect.

IF-THEN The IF-THEN statement is designed to make decisions based on comparisons between two quantities or expressions. It makes use of any of the algebraic comparison relations: $=$, $<$, $>$, $<=$ (less than or equal to), $>=$ (greater than or equal to), and $<>$ (not equal to). The structure of this statement is

[line #]   IF   [expression   relation   expression]   THEN [statement]

The two expressions can be any algebraic expressions at all, and the statement following THEN is any BASIC statement. This latter statement is conditionally executed depending on the truth or falseness of the comparison condition. As examples, you might use

```
200   IF X = 3 THEN PRINT "X IS THREE"
210   IF Y < 5 THEN INPUT "THE NEW VALUE FOR
      Y IS ";Y
```

```
220    IF B^2 - 4 * A * C >= 0 THEN P
       = (B^2 - 4 * A * C)^(.5)
230    IF X + Y <> 14 * P/Q THEN GO TO 200
```

1. The use of the IF-THEN statement provides the way out of the infinite loop in the sample program above. Suppose you insert the line

```
15    IF R = 0 THEN GO TO 999
```

When the condition of the IF statement is fulfilled (if R = 0), then the computer performs the THEN statement (it will go to line 999, leave the loop, and END the program). If the condition is not fulfilled (R has any value other than 0), the computer simply proceeds on to the next line. Therefore, the program will continue to function while you supply different values for the radius R, but will stop as soon as you give R the value 0.

Other uses of the IF-THEN structure are to check for possible division by 0 *before* having the computer perform the division step or to check if a quantity is negative *before* taking its square root.

2. A common variation on the IF-THEN structure is

```
[line #]      IF      ...      THEN      [line #]
```

as in

```
200    IF A = B THEN 500
```

where there is implied a GO TO statement. Another similar variation is

```
[line #]      IF      ...      GO TO      [line #]
```

3. Some versions of BASIC allow the structure

```
[line #]      IF      ...      THEN      ...
ELSE      ...
```

With this, if the condition holds, the computer will perform the THEN statement. If the condition does not hold, it will perform the ELSE statement. In either case, the program will continue to the next line. For example:

```
250    IF D >= 0 THEN S = D^(.5) ELSE PRINT
       "NO SQUARE ROOT "
```

FOR-NEXT One of the most powerful features of a computer is its ability to perform repetitive operations. You can implement this most easily by using a loop construction based on the FOR-NEXT command. It is illustrated in the following examples.

EXAMPLE 1.2

```
100    FOR X = 1 TO 20
110    LET Y = X^2
120    PRINT X,Y
130    NEXT X
140    END
```

Notice that the commands FOR and NEXT encompass a series of statements at lines 110 and 120. The effect here is that the computer starts at line 100 by setting X = 1. It then proceeds on to line 110 to find Y = 1 and then, at line 120, to print the values of x and y (1 and 1) before reaching line 130. At line 130, it resets x to the next value, namely $x = 2$, and then goes back through the loop again, this time using $x = 2$ to calculate $y = 4$. It does this repeatedly until $x = 20$, at which time it passes through the loop to find and print Y = 400. Since there are no more values for x, the computer then leaves the loop and proceeds on to line 140, which ends the program. The result is a table of the squares of the integers from 1 to 20.

EXAMPLE 1.3

```
100    FOR I = 1 TO 100
110    LET S = S + I
120    NEXT I
130    PRINT S
140    END
```

In this program, the index I runs from 1 up to 100. When the program begins, the value for S (as for all variables) is automatically set at 0. Thus, when I = 1 on the first pass through the loop, line 110 produces

```
S = S + I = 0 + 1 = 1
```

on the second pass with I = 2, this becomes

```
S = S + I = 1 + 2 = 3
```

on the third pass with I = 3, this becomes

```
S = S + I = 3 + 3 = 6
```

and so on until I = 100. In effect, then, this program will add up the first 100 integers before leaving the loop and will then print out the sum S at the end.

The index variable in the FOR statement does not have to appear explicitly within the loop.

EXAMPLE 1.4

```
100    FOR N = 1 TO 1000
110    PRINT "THIS IS FUN "
120    NEXT N
130    END
```

In this program, N is used to count off 1000 repetitions of printing the sentence "THIS IS FUN"; it has no other use.

The index variable in a FOR-NEXT loop need not always start at 1 as in the above examples. It could read, for example,

```
FOR N = 5 TO 607      or      FOR X = -7 TO 35
```

In the above cases, the index variable always increased in steps of 1. It is possible to have steps of any size. For instance,

```
FOR X = 0 TO 1000 STEP 20
```

uses the values for $x = 0, 20, 40, 60, \ldots, 1000$, while

```
FOR Y = 5 TO 8 STEP .01
```

uses $y = 5, 5.01, 5.02, \ldots, 8$, and

```
FOR Z = 16.4 TO 8.1 STEP -.1
```

uses $z = 16.4, 16.3, 16.2, \ldots, 8.1$.

BUILT-IN FUNCTIONS The BASIC language contains a series of common mathematical functions built-in as part of the language. Some of the more useful of these are:

SQR = square root

SIN = sine (argument must be in radians)

COS = cosine (argument must be in radians)

TAN = tangent (argument must be in radians)

ATN = arc tangent (result is in radians)

LOG = log to the base e (which will be studied as part of calculus)

EXP = exponential function, powers of e

ABS = absolute value function

In order to use any of these, as well as several other functions, you have to include an *argument* with the function following the structure:

```
FUNCTION(      [expression]      )
```

Thus, for instance, you could write

```
SQR(25), SIN(X)      (X in radians is essential)
TAN(SQR(1 + X^2))
```

and so forth. These might be used in a program as shown in the following example.

EXAMPLE 1.5

```
100   FOR X = 0 TO 100
110   LET S = SQR(X)
120   PRINT X,S
130   NEXT X
140   END
```

This simple program produces a table of square roots of all integers from 0 to 100.

DEF Often in mathematics, we must deal with functions other than the ones built into BASIC. The DEF command allows you to define your own functions. The function to be defined usually must have a name consisting of three letters starting with FN. Thus, the only allowable function names are FNA( ), FNB( ), ..., FNZ( ) where the argument can be any variable desired. The form for the DEF command is

```
100   DEF FNA(X) =     [any expression in X]
```

For example,

```
100   DEF FNA(X) = X * SQR(X^2 + 3)
```

or

```
200   DEF FNB(Q) = SIN(Q)/(Q + 5 * Q^3
      - 17)^(5/3)
```

Once such a function has been defined using the DEF statement, you can use it any time thereafter in a program by simply referring to the function (without the DEF part), using any desired argument. Thus, you could have such commands as

```
250   PRINT FNA(3),FNB(K)
```

or

```
260   IF FNA(P) < FNB(5) THEN PRINT "DONE"
```

INT Several of the built-in functions require special explanations. One of these is INT(X) which is the greatest integer function. For any argument x, INT(X) rounds the argument *down* to the next lowest integer. Thus,

```
INT(5.3) = 5     INT(.024) = 0
INT(-3.7) = -4
```

One application of this function is in the problem of rounding numbers. Suppose you want to round a number x to the nearest integer. You cannot use INT(X), since this rounds down only so that INT(5.8) = 5 rather than 6 as you would want. However, if we increase x by .5, so that 5.8 becomes 6.3, then INT(5.8 + .5) = INT(6.3) = 6. On the other hand, if this is done to 5.1, say, then INT(5.1 + .5) = INT(5.6) = 5. Therefore, INT(X + .5) always rounds x to the nearest integer.

If you want to round x to the nearest hundredth, say, then you have to modify this procedure somewhat. Suppose x were 3.1416. Consider $100x$ = 314.16. You can round this to the nearest integer by adding .5 and taking INT(314.16 + .5) = INT(314.66) = 314. Now divide the result by the same 100 that you used before and obtain INT(100 * X + .5)/100 = 3.14. In general, to round x to n decimal places, you need merely apply

```
INT(10^N * X + .5)/10^N
```

RND Another built-in function is RND that is designed to produce a random number. If you use RND(X) or RND(0) in a program, then the computer will produce a randomly generated number between 0 and 1, say .3859022. Each time it is applied, RND produces a different number.

To produce a random number between 0 and 10, instead of between 0 and 1, all you need to do is multiply RND(X) by 10: 10 * RND(X). Similarly, 100 * RND(X) will produce a random (decimal) number between 0 and 100 and, in general, K * RND(X) is a random number between 0 and k. To get a random (decimal) number between a and b, you would have to use the following scheme: Since (B − A) * RND(X) is a random number between 0 and B − A, it follows that (B − A) * RND(X) + A is a random number between a and b.

Finally, if you need a random whole number, just apply INT in conjunction with RND. Thus,

```
INT((B - A) * RND(X) + A)
```

is a random integer between A and B − 1.

REM REM provides a way of including remarks such as headings and explanations within a program. The computer ignores these lines when RUNning a program, but they can help anyone reading a program LISTing to follow the purpose and logic of the program. You might use something like

```
1   REM THIS PROGRAM SOLVES A QUADRATIC
    EQUATION
```

TWO
THE FUNCTION CONCEPT

2.1
DEFINITION OF A FUNCTION

One of the principal uses of the computer in a calculus course is to perform calculations that might otherwise be too difficult to do by hand. This can be clearly illustrated by considering the definition of a function.

DEFINITION

A *function* is a rule such that for each value of the independent variable, x, in the domain, there corresponds a single value of the dependent variable, y, in the range.

Once we supply the computer with an expression for a particular function, it will respond with the value of y corresponding to any x fed to it. A simple program to accomplish this is shown below.

Program VALUE

```
10  REM VALUE OF A FUNCTION
20  DEF FNY(X) = ...
30  INPUT "ENTER X ";X
40  LET Y = FNY(X)
50  PRINT X,Y
60  END
```

EXAMPLE 2.1

Suppose we consider the rational function

$$f(x) = \frac{x^2 - 9}{x - 3}$$

that is not defined at $x = 3$. In order to apply the program to this function, we have to express it as an appropriate BASIC expression, namely:

```
FNY(X) = (X^2 - 9)/(X - 3)
```

on line 20. (As pointed out in Chapter 1, particular computers may use ** or ↑ in place of ^ for exponents.) If we now input a value of x, say $x = 5.293$ on line 30, the computer will respond with the output

```
5.293    8.293
```

If we now change the value of x at line 30 to $x = 7.456$, say, then we obtain as output

    7.456   10.456

Alternatively, if we supply $x = -18.052$, then the output is

    -18.052   -15.052

In each of these instances, the value for the function is precisely three more than the x value used; that is, the function is actually $x + 3$. We note that this is the result of reducing the original expression:

$$(x^2 - 9)/(x - 3) = (x + 3)(x - 3)/(x - 3) = x + 3$$

which is permitted if $x \neq 3$.

Now consider how the computer would react if we mistakenly try to use the value $x = 3$ at line 30 for this function. That is, how does the computer handle an instruction to perform an undefined operation such as division by zero? The computer responds with a predetermined message indicating an error, perhaps "DIVISION BY ZERO" or "UNDEFINED OPERATION" or something similar (the specific message varies from system to system) and the program STOPs on the error.

EXAMPLE 2.2   We consider a slightly more complicated function, $f(x) = \sqrt{x(x - 2)}$, that is not defined for $0 < x < 2$. We supply this function to the computer at line 20 of the program VALUE using the BASIC expression

    20   DEF FNY(X) = SQR(X * (X - 2))

If we input the value $x = 5$ at line 30, we obtain

    5   3.87298

If we supply $x = -2.822$, then the output is

    -2.822   3.68886

However, if we inadvertently attempt to use a value of x where the function is not defined, say $x = 1$, then the computer responds with an error message such as "IMAGINARY NUMBER" or "SQUARE ROOT OF A NEGATIVE NUMBER."

EXAMPLE 2.3   Suppose we try the function $f(x) = x^4 \sin x$. We supply it as

    20   DEF FNY(X) = X^4 * SIN(X)

If we then supply $x = 1$ at line 30, the output is

    1   .841471

If we want to use $x = \pi/2$, though, we have to input

```
30   LET X = 3.14159/2
```

(Some versions of BASIC allow the use of PI for π, but most do not.) The corresponding output will be

```
1.5708    6.08805
```

Finally, if we try $x = 56.83$, then the result will be

```
56.83    .289592 E7
```

where the answer is automatically converted to exponential form and represents

$$.289592 \times 10^7 \quad \text{or} \quad 2.89592 \times 10^6 \quad \text{or} \quad 2895920 \text{ (rounded off)}$$

EXERCISE 1

Modify the PRINT statement in the program VALUE so the output will look like

```
X = ...      F(X) = ...
```

EXERCISE 2

Modify the program VALUE so it contains a loop in which the value of the function will be calculated for a succession of x values. Use a GO TO command to form the loop. Remember to provide an exit from the loop.

The program VALUE above is intended primarily to introduce the use of the computer, but actually serves little in the way of a practical purpose. The same calculations can be done using a hand-held calculator. However, a further modification of the program makes it extremely useful for generating tables of values for a function similar to the familiar tables for the trigonometric and logarithmic functions. The program is shown below as the program TABLE.

Program TABLE

```
10   REM TABLE OF VALUES FOR A FUNCTION
20   DEF FNY (X) = ...
30   LET A = ...
40   LET B = ...
50   LET N = ...
60   PRINT "X","FNY(X)"
70   FOR X = A TO B STEP (B - A)/N
80   PRINT X,FNY(X)
90   NEXT X
100  END
```

The program TABLE instructs the computer to calculate and print the value of the given function as a succession of $n + 1$ evenly spaced points from a to b. In particular, if we set $h = (b - a)/n$ as the uniform spacing or step between points, then the function is evaluated at the points $x = a$, $x = a + h$, $x = a + 2h$, $x = a + 3h$, ..., $x = a + nh = b$. The corresponding values for the function are FNY(A), FNY(A + H), FNY(A + 2H), ..., FNY(B).

In order to use TABLE, we must supply an expression for the function at line 20, the range of values of x to be considered from a to b at lines 30 and 40, and the number of values, n, of x in the interval that will be used. For example, suppose we choose the function $(x^3 + 5)/\sqrt{x + 2}$ on the interval $[0, 2]$ with $n = 10$ steps. The input lines are then

```
20   DEF FNY(X) = (X^3 + 5)/SQR(X + 2)
30   LET A = 0
40   LET B = 2
50   LET N = 10
```

The corresponding output would then be

```
X          FNY(X)

0          3.53553
.2         3.37639
.4         3.2688
.6         3.23483
.8         3.29405
1          3.4641
1.2        3.76107
1.4        4.19977
1.6        4.79401
1.8        5.5567
2          6.5
```

If we want more detail — that is, a finer subdivision of the interval — then all we need to do is change line 50 to read

```
50   LET N = 40
```

say, and the result would be, in part,

```
X          FNY(X)

0          3.53553
.05        3.49224
.1         3.45102
.15        3.41227
.2         3.37639
.25        3.34375
.3         3.31471
.35        3.28961
.4         3.2688
⋮          ⋮
```

Alternatively, we could change the interval by altering the values for a and b at lines 30 and 40 of the program. In this way, a considerable amount of information about a function can be generated quickly and easily. For example, the above data is useful if we need to graph the given function between $x = 0$ and $x = 2$.

Moreover, we can use this program on functions that otherwise would be far too complicated to consider. For instance, suppose we want to determine a table of values for the function

$$\frac{x^{3/5}10^{\sin x}}{\sqrt[3]{4 + 7x - \tan x}}$$

We merely have to type

```
20   DEF FNY (X) = X^(3/5) * 10^(SIN(X))/
     (4 + 7 * X - TAN(X))^(1/3)
```

If we now apply the program TABLE with $n = 10$ steps across the interval $[a, b] = [0, 1]$, then we obtain the following.

X	FNY(X)
0	0
.1	.190075
.2	.347292
.3	.534011
.4	.762859
.5	1.04248
.6	1.37892
.7	1.77503
.8	2.22951
.9	2.73623
1.0	3.28428

A plot of these points is shown in Figure 2.1.

While this procedure to draw the graph of a function may seem very easy and appealing, it has some possible drawbacks. By simply connecting the points with a smooth curve as we did in Figure 2.1, we implicitly assume that the graph of the function is relatively smooth. This assumption can be wrong. A particular function may have some totally unexpected behavior between two successive points calculated by the computer. Moreover, the use of a program such as TABLE automatically restricts attention to the particular range of values selected. Often, interesting aspects of the behavior of a function can occur well outside such an interval and therefore would go unnoticed. A much safer strategy is to use some of the techniques of calculus to be developed later in the course that are designed to characterize the precise behavior of a function. Nevertheless, if used with caution, a program such as TABLE can be a valuable tool for obtaining some initial information about a function.

FIGURE 2.1

Graph of $\dfrac{x^{3/5}10^{\sin x}}{\sqrt[3]{4 + 7x - \tan x}}$

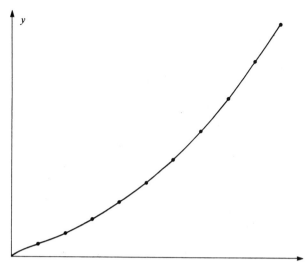

Note: x ranges from 0 to 1 in steps of .1
y ranges from 0 to 3.28428 in steps of .328428

EXERCISE 3

Modify program TABLE to include appropriate headings for each column of the output.

EXERCISE 4

Modify program TABLE by replacing the LET statements at lines 30, 40, and 50 with INPUT statements. Have each INPUT statement be accompanied by an appropriate message or prompt to indicate which quantity is being supplied.

SECTION 2.1 PROBLEMS

Use the program VALUE or one of your modifications of it to calculate the value of each of the following functions at the indicated points.

1 $f(x) = \dfrac{x^3 - 2x^2}{x - 2}$ at $x = 0, 1, 1.5, 1.9, 1.99, 1.999$

2 $f(x) = \sqrt{\dfrac{x^3 - 2x^2}{x - 2}}$ at $x = 0, 1, 1.5, 1.9, 1.99$

3 $f(x) = \sqrt{x^2 - 3x - 4}$ at $x = 5, 10, 15, 20$

4 $f(x) = \dfrac{x^3 + 5x^2 - 6x - 30}{x + 5}$ at $x = 3, -3, -6$

5 $f(x) = x^{\sin x}$ at $x = 1, 2, 3, 3.14$

Use the program TABLE or your modification of it to generate a table of values for each of the following functions using the indicated intervals and the specified number of points.

6 $f(x) = \sin(\pi x/180)$ on $[0, 90]$, $n = 18$. (Note: The argument $\pi x/180$ is used instead of simply x to convert from degrees to radians.)

7 $f(x) = \cos(\pi x/180)$ on $[0, 90]$, $n = 90$

8 $f(x) = \sqrt{x}$ on $[0, 25]$, $n = 25$

9 $f(x) = 2^x$ on $[0, 20]$, $n = 20$

10 $f(x) = \dfrac{x^3 - 2x^2}{x - 2}$ on $[-1, 1]$, $n = 20$

11 $f(x) = \sqrt{x^2 - 3x - 4}$ on $[5, 100]$, $n = 19$

12 $f(x) = |x||x - 1|$ on $[-3, 5]$, $n = 16$

2.2
BEHAVIOR OF FUNCTIONS

In the previous section, we considered ways in which the computer can be used to supply information about the values of a function. Essentially, this is a purely algebraic process. In calculus, however, we are more concerned with the behavior of a function—the rate at which its values change—rather than with the values themselves. Nevertheless, some information about behavior can be determined by examining and comparing the values of a function at nearby points.

For example, consider the function $x^3 - 3x^2 + 2$ on the interval $[-1, 3]$. We apply the program TABLE to it using $n = 10$ and obtain the output shown in Table 2.1. Based on this data, we may conclude that the function increases in value from $x = -1$ to $x = .2$, then decreases in value from $x = .2$ to $x = 1.8$, and finally again increases from $x = 1.8$ to $x = 3$. However, the behavior of the function could change dramatically within any of these subintervals. Moreover, it is highly unlikely that the function would increase until $x = .2$ exactly and then decrease from $x = .2$ to 1.8 exactly, and so on. Rather, it is likely that the initial increase occurs

TABLE 2.1

Table of values of $f(x) = x^3 - 3x^2 + 2$ on $[-1, 3]$

X	FNY(X)
-1	-2
-.6	.704
-.2	1.872
.2	1.888
.6	1.136
1	0
1.4	-1.136
1.8	-1.888
2.2	-1.872
2.6	-.704
3	2

from $x = -1$ to somewhere near $x = .2$, and so forth. Therefore, to obtain a more accurate picture of this function's behavior, we can run the program again with a larger number of points, say $n = 20$. The corresponding result is presented in Table 2.2. When we examine these values, we see that our original "conclusions" have to be modified. It now appears that the function increases from $x = -1$ to $x = 0$, then decreases from $x = 0$ to $x = 2$, and then increases from $x = 2$ to $x = 3$. Again, though not conclusive, this is more accurate than the analysis based on the first set of results. If the value for n were increased still further, say to $n = 100$ or even $n = 1000$, then a far more accurate picture of the function's behavior would emerge. Unfortunately, while such a comprehensive set of values would give extremely accurate information about the function, it would also be a tremendous amount of data to analyze by hand. Therefore, it makes sense to incorporate such an analysis into the computer program and have the computer produce only the pertinent conclusions, but not the mass of intermediate values.

The resulting program is fairly sophisticated and therefore beyond the skills of most beginning programmers at this stage. For this reason, the program is available as a canned program under the name INCDEC (for INCreasing-DECreasing). Your instructor will supply details on how this program can be called on by the students at each installation. Instructions for using the program are incorporated within it.

TABLE 2.2

More detailed table of values for $f(x) = x^3 - 3x^2 + 2$

X	FNY(X)
-1	-2
-.8	-.432
-.6	.704
-.4	1.456
-.2	1.872
0	2
.2	1.888
.4	1.584
.6	1.136
.8	.592
1	0
1.2	-.592
1.4	-1.136
1.6	-1.584
1.8	-1.888
2	-2
2.2	-1.872
2.4	-1.456
2.6	-.704
2.8	.432
3	2

SECTION 2.2 PROBLEMS

Use the canned program INCDEC to analyze the behavior of each of the following functions. In each case, select what you consider an appropriate interval $[a, b]$ and an appropriate step h. Change the values of these parameters to see the effects.

1 $f(x) = 2x^3 - 6x + 1$ **2** $f(x) = x^3 + x^2 - 5x$

3 $f(x) = 3x^4 + 2x^2$ **4** $f(x) = x^4 - 3x^3 + 3x^2 + 1$

5 $f(x) = 5x^{2/3} - x^{5/3}$ **6** $f(x) = x^2(x + 4)^3$

7 $f(x) = x^{1/3} + 2x^{4/3}$

8 $f(x) = x^6 - 29x^5 - .2x^4 + x^3 - 7x^2 + 5$

9 $f(x) = 3 + (x + 1)^{7/5}$ **10** $f(x) = x^2\sqrt{4 - x}$

11 $f(x) = \dfrac{(x + 1)^2}{x^2 + 1}$ **12** $f(x) = \dfrac{x - 1}{x^2 - 2x + 2}$

13 $f(x) = \dfrac{9x}{x^2 + 9}$ **14** $f(x) = (x + 2)\sqrt{-x}$

15 $f(x) = \sin \dfrac{x^2}{x + 5}$ **16** $f(x) = x^2 \sin \dfrac{10}{x}$

17 $f(x) = x^{\cos x} \ (x \geq 0)$

2.3
SEQUENCES

We now consider a concept fundamental to any use of the computer. A *sequence* is any set of real numbers in a particular order. That is, there is a first number a_1, a second a_2, a third a_3, and so forth. We denote the general or n^{th} term of a sequence by a_n. The entire sequence is expressed as

$$A = \{a_1, a_2, \ldots, a_n, \ldots\}$$

Furthermore, as a useful shorthand notation, we express the sequence A as the general term $A = \{a_n\}$.

As examples of sequences, we have

$$A = \{2, 4, 6, 8, 10, \ldots\}$$

$$B = \{2, 4, 8, 16, 32, \ldots\}$$

$$C = \left\{\frac{1}{2}, \frac{2}{4}, \frac{3}{8}, \frac{4}{16}, \frac{5}{32}, \ldots\right\}$$

In each of these examples, it is simple to determine the patterns from which we can supply meaning to the dots by predicting the following terms. In particular, this is done by obtaining a formula for the general term in each. Thus, for each value of $n = 1, 2, 3, \ldots$, we have

$$A = \{a_n\} = \{2n\}$$

$$B = \{b_n\} = \{2^n\}$$

$$C = \{c_n\} = \left\{\frac{n}{2^n}\right\}$$

A more formal definition of a sequence follows.

DEFINITION

A *sequence* is a function with domain the set of all positive integers and with range in the set of all reals.

In other words, for each positive integer $n = 1, 2, 3, \ldots$, there corresponds a real number a_n as the nth term in the sequence. This is illustrated by the general term found for the above examples. For example, in sequence A, the first term is $a_1 = 2$, the second is $a_2 = 4$, and so forth.

Furthermore, we note that the above definition can be extended to have a sequence defined on the non-negative integers. This allows starting with a 0th term, a_0, for a sequence $\{a_0, a_1, a_2, \ldots\}$. This latter possibility is often useful and convenient.

In all of the above, we have tacitly assumed that all sequences continue indefinitely; that is, they are *infinite sequences*. This is not necessarily the case. Often, one deals with sequences that have only a fixed finite number of elements. A classic example is the sequence

$$\{4, 14, 34, 42, 47, 59, 125, \ldots, 204\}$$

that cannot be expressed by any simple mathematical formula. In fact, the numbers represent the stops on one of the New York City subway lines.

In virtually all cases we will consider, the sequences used will be nonterminating. However, in a practical sense, we can deal only with a finite number of terms of such a sequence. Thus, we will primarily be concerned with the first k terms of an infinite sequence for some positive integer k.

EXERCISE 5

Prepare a program that calculates and prints out the first k terms of any sequence. In particular, supply the value for k and the expression for the general term of the sequence A in terms of n using a LET statement. Use a FOR-NEXT loop to perform the work.

If we apply a program such as the one in Exercise 5 to the sequence $A = \{(n - 1)/n\}$, then we obtain

```
1        0
2        .5
3        .666667
4        .75
```

5	.8
6	.833333
7	.857143
8	.875
9	.888889
10	.9
11	.909091
12	.916667
13	.923077
14	.928571
15	.933333

Alternatively, if we use the sequence

$$A = \left\{ (-1)^n \frac{n^2}{5^n} \right\}$$

then we find

1	-.2
2	.16
3	-.072
4	.0256
5	-.008
6	.002304
7	-.006272
8	.0001638

We now consider some additional properties of sequences. However, rather than concentrating on the first k terms as we have done so far, we will study instead the behavior of terms towards the extreme end of the sequence. To begin, we again examine the sequence

$$A = \left\{ \frac{n-1}{n} \right\} = \left\{ 0, \frac{1}{2}, \frac{2}{3}, \frac{3}{4}, \frac{4}{5}, \cdots \right\}$$

considered above. Obviously, every term of this will be a fraction, but consider what happens to these fractions for large values of n. For example, suppose $n = 100$, so that the corresponding term $a_{100} = .99$, which is quite close to 1. What if n were much larger, say $n = 1,000,000$? The corresponding element of the sequence would be

$$a_{1000000} = \frac{999,999}{1,000,000} = .999999$$

which is certainly much closer to 1 than a_{100} is. In this way, it is clear that, as n gets larger and larger, the values for a_n get progressively closer to 1. We speak of the value 1 as the *limit* of this sequence in the sense that it is the limiting value for the elements of the sequence—they get closer and closer to it the further out in the sequence we go.

However, this limiting value is *never* actually reached. In the above example, there is no value for n, no matter how large, for which $(n - 1)/n$

is exactly equal to 1. On the other hand, if we take a large enough value for n, we can get a term a_n as close to 1 as we wish.

EXAMPLE 2.4

Consider the sequence with $a_n = 1/2^n$, for $n = 1, 2, 3, \ldots$. As n gets larger and larger, 2^n becomes increasingly larger, so that the terms in the sequence

$$\left\{ \frac{1}{2}, \frac{1}{4}, \frac{1}{8}, \frac{1}{16}, \frac{1}{32}, \frac{1}{64}, \frac{1}{128}, \frac{1}{256}, \frac{1}{512}, \cdots \right\}$$

$$= \{.5, .25, .125, .0625, .03125, .015625, \ldots\}$$

become smaller and smaller and consequently approach a limiting value of 0.

EXAMPLE 2.5

Consider the sequence with general term $a_n = n^2$ for $n = 1, 2, 3, \ldots$. As n gets larger, the corresponding term a_n also becomes infinitely large and consequently does not approach any finite limiting value, so that no limit exists for this sequence.

In the example above, the term "infinitely large" was used with the understanding that it is a purely intuitive concept that cannot be defined. We now formulate a more precise definition of the concept of limit of a sequence.

DEFINITION

A number L is called the *limit* (*limiting value*) of a sequence a_n if, as n becomes infinitely large, all of the corresponding terms a_n approach arbitrarily close to L. This is written

$$\lim_{n \to \infty} a_n = L$$

Mathematically, this means that if we select any open interval about L, no matter how small, then from some particular value of n on, call it N, all terms of the sequence will lie within the interval. That is, if we want the terms of the sequence to be within a given tolerance E (usually quite small) of the limit L, then we will have

$$|a_n - L| < E$$

whenever $n \geq N$, for some N. A sequence whose limit L exists is said to be *convergent* or to *converge to L*. If no such limit exists, then the sequence is said to be *divergent*.

In the above examples, the sequence $\{1/2^n\}$ converges since the limit $L = 0$ exists, while the sequence $\{n^2\}$ diverges since it does not possess a limit.

It is essential to realize that although the terms a_n of a convergent sequence approach arbitrarily close to the limit L as $n \to \infty$, none of the a_n will be exactly equal to L except in certain unusual circumstances.

EXAMPLE 2.6

Consider the sequence $\{1, \frac{1}{2}, 1, \frac{2}{3}, 1, \frac{3}{4}, 1, \frac{4}{5}, \ldots\}$. The limit of the sequence is certainly $L = 1$. However, every second term in the sequence is already equal to the limit.

EXAMPLE 2.7

Consider the sequence with $a_n = 3 + (5/n)$, for $n = 1, 2, 3, \ldots$. As n becomes large, clearly $1/n$ approaches zero and consequently, $5(1/n) = 5/n$ also approaches 0. Therefore, as n approaches infinity, the term $3 + (5/n)$ becomes ever closer to $3 + 0 = 3$, so that

$$\lim_{n \to \infty}\left(3 + \frac{5}{n}\right) = 3 = L$$

and this sequence is convergent.

EXAMPLE 2.8

Consider the sequence $\{1, 0, 1, 0, 1, 0, \ldots\}$ in which every odd-numbered term is equal to 1 and every even-numbered term is equal to 0. As n approaches infinity, the terms continually alternate between 0 and 1, but do not *all* approach any specific limit L. (Half of them approach, and are in fact equal to, 0; the other half approach 1.) Thus, this sequence diverges (no limit exists).

EXERCISE 6

Write a program designed to find the limit of a given sequence. In particular, supply the general term of any sequence in terms of N (DEF FNS(N) = . . .). Have the computer calculate and compare successive values of the sequence. If a pair of successive terms are virtually equal (for example, if the absolute value of their difference is almost zero, ABS(FNS(N) − FNS(N + 1)) = .00001, say), interpret this to mean that the limit essentially has been reached and have the computer print out an appropriate message. Include a maximum number of terms to be checked in case no limit is reached within a reasonable time, and print out an appropriate message in such an instance.

You should realize that the program suggested in Exercise 6 can be quite misleading and should be used only to obtain an intuitive feel for the concept of limit of a sequence. For example, if it were used on the sequence $\{1, 5, 2, 5, 3, 5, 4, 5, 5, 5, 6, 5, 7, 5, \ldots\}$, the program would find a limit of 5, since several successive terms are equal, but actually the sequence diverges.

Finally, note that in many of the computer applications that follow, we will use some sequences that converge to 0. We already saw several examples of such sequences, such as $\{1/n\}$ and $\{1/2^n\}$. For the sake of comparison, we examine the first few terms of each:

$$\left\{\frac{1}{n}\right\} = \{1, .5, .333333, .25, .2, \ldots\}$$

$$\left\{\frac{1}{2^n}\right\} = \{.5, .25, .125, .0625, .03125, \ldots\}$$

Clearly, the terms in $\{1/2^n\}$ converge to 0 at a much faster rate than do those in $\{1/n\}$. In fact, the convergence of $\{1/n\}$ is so slow that it is practically useless to utilize it in any calculation — far too many terms would have to be considered. Even the similar sequence

$$\left\{\frac{1}{n^2}\right\} = \{1, .25, .111111, \ldots\}$$

converges too slowly for most purposes. Thus, in order to get a reasonable set of values in most cases, we need to use sequences that converge quickly. A good example is the sequence

$$\left\{\frac{1}{10^n}\right\} = \{1, .1, .01, .001, .0001, \ldots\}$$

However, there are situations where the convergence of this sequence is too rapid, and the terms get too close to zero before we can obtain any significant information. For these reasons, no specific sequence is suggested as being the most useful. In some situations, a sequence such as $\{1/n^2\}$ or $\{1/n^3\}$ or $\{1/2^n\}$ that converges moderately quickly is most informative. Often, a sequence such as $\{1/10^n\}$ that converges extremely rapidly is the best choice. Rarely, a sequence that converges much faster, such as $\{1/20^n\}$, is the preferred choice.

However, by actually trying a variety of different sequences converging at different rates, we can develop a much deeper insight into and understanding of the processes we are studying.

SECTION 2.3 PROBLEMS

Write out the first six terms of the following sequences whose general term is given.

1 $a_n = 4n$

2 $a_n = 3n + 5$

3 $a_n = \dfrac{1}{2^n}$

4 $a_n = n^2 + 5$

5 $a_n = n^3 - 10$

6 $a_n = \dfrac{n^2 + 1}{n^2 + 2}$

7 $a_n = \dfrac{2^n}{3^n}$

8 $a_n = \dfrac{n^2}{2^n}$

Determine an expression for the general term in each of the following sequences and use it to predict the next two terms.

9 $\{3, 5, 7, 9, 11, \ldots\}$ **10** $\{2, 5, 8, 11, 14, 17, \ldots\}$

11 $\{192, 96, 48, 24, 12, \ldots\}$ **12** $\{2, 5, 10, 17, 26, 37, 50, 65, 82, \ldots\}$

13 $\{1/3, 2/4, 3/5, 4/6, \ldots\}$ **14** $\{2/5, 4/25, 8/125, \ldots\}$

Use the program suggested in Exercise 6 to find the approximate limit for any of the following sequences that converge. Verify the results by applying the concepts on limits of sequences.

15 $\left\{\dfrac{1}{3n}\right\}$ **16** $\left\{\dfrac{1}{3^n}\right\}$

17 $\left\{\dfrac{3^n}{2^n}\right\}$ **18** $\left\{\dfrac{n}{n^2 + 1}\right\}$

19 $\left\{1 + \dfrac{n}{n + 1}\right\}$ **20** $\left\{\dfrac{2n + 1}{n + 1}\right\}$

21 $\left\{\left(\dfrac{2}{3}\right)^n\right\}$ **22** $\{(-1)^n\}$

23 $\left\{\dfrac{(-1)^n}{n}\right\}$ **24** $\left\{\dfrac{\sin n}{n}\right\}$

25 $\{\cos n\}$ **26** $\left\{\dfrac{n^2 - 5}{n^2 + 5}\right\}$

27 $\left\{\dfrac{n}{2^n}\right\}$ **28** $\left\{\dfrac{n^2}{2^n}\right\}$

29 $\left\{\dfrac{2^n}{n^5}\right\}$ **30** $\left\{\dfrac{3n^2 + 1}{4n^2 + 2}\right\}$

2.4
LIMIT OF A FUNCTION

In the present section, we use the computer to illustrate the concept of the limit of a function at a point. That is,

$$\lim_{x \to a} f(x) = L$$

means that as x approaches a, the values of the function $f(x)$ become arbitrarily close to a fixed limiting value L. If the definition were applied rigorously, it would require that we check what happens to the values of the function along *every* possible way that x can approach a. Clearly, this involves an unlimited number of cases and as such cannot be done practically. Rather, we will select several typical routes by which x approaches a from either side to gain a feel for what happens. However, you must understand that the results will not constitute formal proof of the existence of a limit.

Consider the function

$$f(x) = \frac{2x^2 + x - 3}{x - 1}$$

that is well-defined for all values of x except $x = 1$. In fact, so long as $x \neq 1$, we can perform the division $(2x + 3)(x - 1)/(x - 1) = 2x + 3$ and thus reduce the expression to

$$f(x) = 2x + 3, \qquad x \neq 1$$

Thus, other than at $x = 1$, the function represents the equation of a line. The only uncertainty about the behavior of this function occurs when x is close to, but not quite equal to, 1, as shown in Figure 2.2. We therefore investigate the behavior of this function as x approaches the value $a = 1$.

To get an idea of what happens close to $a = 1$, we select a sequence of x values which approach $a = 1$. One such sequence is $\{1 + 1/10^n\}$ whose terms are $\{1.1, 1.01, 1.001, 1.0001, 1.00001, \ldots\}$. The corresponding values of the function are:

X	FNY(X)
1.1	5.2
1.01	5.02
1.001	5.002
1.0001	5.0002
1.00001	5.00002
1.000001	5.000002
⋮	⋮

FIGURE 2.2

Graph of $y = \dfrac{2x^2 + x - 3}{x - 1}$

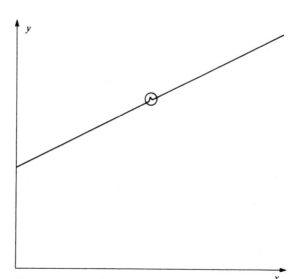

Note: x ranges from 0 to 2 in steps of .2
y ranges from 0 to 7 in steps of .7

Moreover, if we take the sequence of x values $\{1 - 1/10^n\}$ that approach $a = 1$ from the left, we find

X	FNY(X)
.9	4.8
.99	4.98
.999	4.998
.9999	4.9998
.99999	4.99998
.999999	4.999998
⋮	⋮

On the basis of these results, it seems reasonable to conclude that the limit of the function is 5 as x approaches $a = 1$. We note though that x only approaches 1, but never reaches it. Similarly, the values of the function never reach 5.

However, the above analysis is based on just two particular sequences of x values that converge to $a = 1$. We have no guarantee that the same limiting value of 5 would be achieved if we used some other sequences. Suppose, then, that we try the sequence $\{1 + 1/2^n\}$ with terms $\{1.5, 1.25, 1.125, 1.0625, 1.03125, \ldots\}$. The corresponding set of values is

X	FNY(X)
1.5	6
1.25	5.5
1.125	5.25
1.0625	5.125
1.03125	5.0625
⋮	⋮

Again, this suggests that the limit of this function is 5 as x approaches 1, although the sequence used converges at a slower rate than the earlier ones did. It is worth noting, incidentally, that if the sequence of x values converges relatively slowly, then the corresponding sequences of functional values will also converge more slowly (if they converge at all).

Unfortunately, even with the aid of a computer, it is impossible to check more than a finite number of terms of such a sequence. Worse still, it is impossible to check more than a small number of possible sequences that converge to a. The best we can do is check several representative sequences to get a feel for what seems to be happening in the vicinity of the number a.

We now present a program, LIMFN, that can be used to test for the possible existence of the limit for a function. It is based on the use of two sequences of x values, each converging to the value a from different sides.

Program LIMFN

```
10   REM LIMIT OF A FUNCTION
20   DEF FNY(X) = ...
30   INPUT "THE VALUE OF A IS ";A
```

```
40   INPUT "THE NUMBER OF TERMS IS ";K
50   PRINT "A - H ","FNY(A - H) ","A + H ",
     "FNY(A + H) "
60   FOR N = 1 TO K
70   LET H = ...
80   PRINT A - H,FNY(A - H),A + H,FNY(A + H)
90   NEXT N
100  END
```

To use this program, you must supply the following:

1. The desired function FNY(X) in terms of x at line 20;

2. A sequence of numbers h in terms of n that converges to 0 relatively rapidly; for example,

```
70   LET H = 1/10^N
```

3. The number k of terms you want checked in the sequence.

To illustrate the use of this program, suppose we wish to investigate the behavior of the function

$$f(x) = \frac{x^3 + 8}{x + 2}$$

in the vicinity of the point $a = -2$. We first define the function as

```
20   DEF FNY(X) = (X^3 + 8)/(X + 2)
```

For the sequence of numbers that converges to 0, we may select $h = 1/n^2$ for $n = 1, 2, 3, \ldots$. We then type

```
40   LET H = 1/N^2
```

The computer uses this sequence to form a pair of related sequences of x values that approach $a = -2$ from either side. These are the sequences

$$a + h = \left\{-2 + \frac{1}{n^2}\right\} \quad \text{and} \quad a - h = \left\{-2 - \frac{1}{n^2}\right\}$$

Finally, we have to supply the maximum value of n (equivalently, the number of terms of the sequences to be checked), and suppose we choose $k = 15$. The resulting output is presented in Table 2.3.

Based on these results, we suspect that the limit of the function is 12 as x approaches $a = -2$. However, greater accuracy can be achieved by using either a larger value for k or a different sequence for h that converges to 0 at a faster rate. For instance, we may try $h = 1/2^n$, so that the sequences of x values used are

$$a + h = \left\{-2 + \frac{1}{2^n}\right\} \quad \text{and} \quad a - h = \left\{-2 - \frac{1}{2^n}\right\}$$

The resulting table of values from the program LIMFN is presented in Table 2.4.

TABLE 2.3

Table of values for the limit of $f(x) = (x^3 + 8)/(x + 2)$ as $x \rightarrow -2$

A - H	FNY(A - H)	A + H	FNY(A + H)
-3	19	-1	7
-2.25	13.5625	-1.75	10.5625
-2.11111	12.679	-1.88889	11.3457
-2.0625	12.3789	-1.9375	11.6289
-2.04	12.2416	-1.96	11.7616
-2.02778	12.1674	-1.97222	11.8341
-2.02041	12.1229	-1.97959	11.878
-2.01563	12.094	-1.98438	11.9065
-2.01235	12.0742	-1.98765	11.9261
-2.01	12.0601	-1.99	11.9401
-2.00826	12.0497	-1.99174	11.9505
-2.00694	12.0417	-1.99306	11.9584
-2.00592	12.0355	-1.99408	11.9645
-2.0051	12.0306	-1.9949	11.9694
-2.00444	12.0267	-1.99556	11.9734

TABLE 2.4

Another set of values for the limit of $f(x) = (x^3 + 8)/(x + 2)$ as $x \rightarrow -2$

A - H	FNY(A - H)	A + H	FNY(A + H)
-2.5	15.25	-1.5	9.25
-2.25	13.5625	-1.75	10.5625
-2.125	12.7656	-1.875	11.2656
-2.0625	12.3789	-1.9375	11.6289
-2.03125	12.1885	-1.96875	11.8135
-2.01563	12.094	-1.98438	11.9065
-2.00781	12.0469	-1.99219	11.9532
-2.00391	12.0235	-1.99609	11.9766
-2.00195	12.0117	-1.99805	11.9883
-2.00098	12.0059	-1.99902	11.9941
-2.00049	12.0029	-1.99951	11.9971
-2.00024	12.0015	-1.99976	11.9985
-2.00012	12.0007	-1.99988	11.9993
-2.00006	12.0004	-1.99994	11.9996
-2.00003	12.0002	-1.99997	11.9998

If we use a sequence for h that converges to 0 at an even faster rate, such as $h = 1/10^n$, then we would obtain the data shown in Table 2.5.

Again, both of these sets of values seem to indicate that the limit of the function is 12 as x approaches $a = -2$. See Figure 2.3.

It is worth noting, though, that the last set of results is extremely dependent on the computer used. In Section 2.3 on sequences, the point was made that a sequence such as $h = 1/10^n$ may sometimes converge too rapidly and so provide little or no usable information. If the identical program is run on a different computer, the effect of this too-rapid convergence becomes dramatically clear. The new output is shown in Table 2.6. If we

TABLE 2.5

Values for the limit of $f(x) = (x^3 + 8)/(x + 2)$ using $h = 1/10^n$

A - H	FNY(A - H)	A + H	FNY(A + H)
-2.1	12.61	-1.9	11.41
-2.01	12.0601	-1.99	11.9401
-2.001	12.006	-1.999	11.994
-2.0001	12.0006	-1.9999	11.9994
-2.00001	12.0001	-1.99999	11.9999
-2	12	-2	12
-2	12	-2	12
-2	12	-2	12

FIGURE 2.3

Graph of $y = \dfrac{x^3 + 8}{x + 2}$

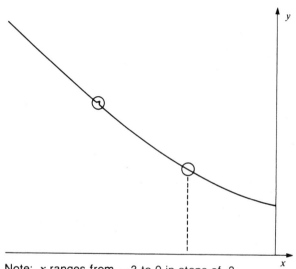

Note: x ranges from -3 to 0 in steps of .3
y ranges from 0 to 19 in steps of 1.9

TABLE 2.6

Alternate values for the limit of $f(x) = (x^3 + 8)/(x + 2)$ using $h = 1/10^n$

A - H	FNY(A - H)	A + H	FNY(A + H)
-2.1	12.610000	-1.9	11.409999
-2.01	12.060100	-1.99	11.940099
-2.001	12.006009	-1.999	11.994000
-2.0001	12.000633	-1.9999	11.999385
-2.00001	12.001490	-1.99999	11.999069
-2.000001	12.003724	-1.999999	11.990685◀
-2.0000001	6.093458	-1.9999999	11.925581
-2	16	-2	8

DIVISION BY ZERO

examine this table carefully starting at the arrow, we see that the values begin to diverge and soon we are given an error message about DIVISION BY ZERO. The problem here involves the way each computer handles small numbers, particularly those that are close to 0/0. At some point, each computer can no longer distinguish the values for the numerator or denominator and so produces misleading round-off errors and/or apparent division by zero. In fact, if we use a sequence for h that converges to zero at a faster rate, say $h = 1/20^n$, then the problem is greatly magnified and occurs much faster. Thus, as was cautioned in the last section, sequences that converge too rapidly are rarely used.

In view of the above example and comments, it should be evident to you that you must take great care in interpreting the computer's output, particularly when computations such as the ones above are being done.

EXERCISE 7

Modify the program LIMFN to compare the values of FNY(A − H) and FNY(A + H) at each stage. In particular, if they are almost equal (say, if ABS(FNY(A + H) − FNY(A − H)) < .0001), have the computer print out an appropriate message regarding the apparent limit of the function.

It is important to realize that the results of such a program can be misleading or even wrong. For instance, consider the function $f(x) = 1/x^2$ as x approaches 0. Clearly, the function is not defined at the origin and hence the limit does not exist. However, if we apply this program to it, we find that, for any $h \neq 0$,

$$\left| f(0 + h) - f(0 - h) \right| = \left| \frac{1}{h^2} - \frac{1}{(-h)^2} \right| = 0$$

and so the program indicates that the limit indeed exists and is equal to 0. As a result, it is essential that you not accept without thought the program's determination of the existence and value of a limit. Rather, you should use it as a preliminary indicator regarding the limit, but one that you must supplement with an examination of actual values of the sequence to see if the decision is reasonable.

EXERCISE 8

Modify the program LIMFN to test if the function values FNY(A + H) and FNY(A − H) are diverging from each other. Specifically, compare the difference D = ABS(FNY(A + H) − FNY(A − H)) at any stage to the corresponding difference at the following iteration. If the difference increases, then the values are apparently diverging.

Once again, we must realize that the computer's decision should not be considered more than a tentative indication. It is certainly possible to construct examples of sequences in which the difference D initially tends to zero and then diverges. Therefore, the computation still must be supplemented with an examination of actual terms of the sequence before a solid conclusion can be drawn.

In summary, the computer's results cannot constitute a rigorous proof of the existence of a limit or of the value of the limit when it exists. At best, they provide a strong indication about the behavior of a function in the limit as well as the numerical data that can lend conviction to the mathematical conclusions.

We will encounter several related programs in later sections involving other applications of limits. In each case, it is important to keep the above cautions in mind, since they apply in all such instances.

SECTION 2.4 PROBLEMS

Apply the program LIMFN or one of your modifications of it to each of the following functions.

1 $\lim\limits_{x \to 2}(x^2 + 2x - 1)$ 　　　　　　　**2** $\lim\limits_{x \to -1}(x^3 - 2x^2 + 3x - 4)$

3 $\lim\limits_{x \to 2}\dfrac{x^2 - 5}{2x^3 + 6}$ 　　　　　　　　**4** $\lim\limits_{x \to -1}\dfrac{2x + 1}{x^2 - 3x + 4}$

5 $\lim\limits_{x \to -2}\dfrac{x^3 + 8}{x + 2}$ 　　　　　　　　**6** $\lim\limits_{x \to 1}\dfrac{x^3 - 1}{x - 1}$

7 $\lim\limits_{x \to -3}\dfrac{x^2 + 5x + 6}{x^2 - x - 12}$ 　　　　　**8** $\lim\limits_{x \to 4}\dfrac{3x^2 - 17x + 20}{4x^2 - 25x + 36}$

Apply the program LIMFN or one of the modifications of it to determine the behavior of each of the following functions near the indicated point.

9 $f(x) = \sqrt{\dfrac{8x + 1}{x + 3}}$ near $x = 1$

10 $f(x) = \sqrt{\dfrac{x^2 + 3x + 4}{x^3 + 1}}$ near $x = 2$

11 $f(x) = \dfrac{2 - \sqrt{4 - x}}{x}$ near $x = 0$ 　　**12** $f(x) = \dfrac{\sqrt[3]{x + 1} - 1}{x}$ near $x = 0$

13 $f(x) = \dfrac{\sqrt{x + 9} - 3}{x}$ near $x = 0$ 　　**14** $f(x) = \sqrt{\dfrac{8x^3 - 27}{4x^2 - 9}}$ near $x = \dfrac{3}{2}$

15 $f(x) = \dfrac{\sin x}{x}$ near $x = 0$ 　　　　　**16** $f(x) = \dfrac{1 - \cos x}{x}$ near $x = 0$

17 $f(x) = \dfrac{10^x - 1}{x}$ near $x = 0$ 　　　　**18** $f(x) = \dfrac{2^x - 1}{x}$ near $x = 0$

19 $f(x) = (1 + x)^{1/x}$ near $x = 0$ 　　　**20** $f(x) = (1 + 1/x)^x$ near $x = 0$

21 $f(x) = (1 + 1/x^2)^{x^2}$ near $x = 0$ 　　**22** $f(x) = (\sin x)^x$ near $x = 0^+$

23 $f(x) = x^{\sin x}$ near $x = 0^+$ 　　　　　**24** $f(x) = (1 - \cos x)^x$ near $x = 0$

25 $f(x) = (1 - \cos x)^{\sin x}$ near $x = 0$

2.5
LIMITS TO INFINITY

In the previous section, we considered how the computer can be used to deal with the limit of a function $y = f(x)$ as x approaches some limit point a. We saw that it was necessary to consider a sequence of numbers $\{a + h\}$ that converges to the value a as h goes to 0. We also saw that it was important to choose a sequence h that approaches 0 at a moderately rapid rate — if it converges too slowly, too much information is produced; if it converges too rapidly, the values calculated for the function will begin to diverge too rapidly for any useful information to be deduced, due to the round-off errors involved.

In the present section, we will take up the notion of the limit to infinity of a function $y = f(x)$. That is,

$$\lim_{x \to \infty} f(x)$$

In view of what has been done before, it is fairly obvious that we should select sequences of values for x that diverge to infinity and then examine the corresponding values for $f(x)$ to decide if a limit exists. As with the previous section, however, it is important to select sequences of x values that approach infinity at a moderately rapid rate, as seen in the example below.

EXAMPLE 2.9   Consider

$$\lim_{x \to \infty} \frac{x^2 + 1}{x^2 + 5}$$

We know that the limit of this function as x approaches ∞ is 1. We will use the computer to verify this result. First, suppose we select the sequence $x = \{n^2\}$. The corresponding sequence of values is shown in Table 2.7. From these values, it certainly appears that the limit is 1. By way of comparison, we have also used two other divergent sequences of x values, $x = \{2^n\}$ and $x = \{10^n\}$. In each case, clearly the limit should be 1.

We should be aware of several important cautions before using the procedure in the example. First, each computer has a built-in limitation on the size of the numbers it can handle. Some models cannot deal with any number exceeding 10^{39}; others cannot accept anything larger than 10^{78}; and so on. When an arithmetic operation leads to a number exceeding the innate capacity of a computer, an overflow condition occurs and the computer automatically stops processing the program at that point. Therefore, we must be careful when using a sequence of values for x that diverges not to carry the calculations too far or we will hit an overflow error. This is what occurred in the last two columns in Table 2.7. Further, if x diverges too rapidly, then

TABLE 2.7

Values for $\lim_{x \to \infty} (x^2 + 1)/$ $(x^2 + 5)$

n	$f(n^2)$	$f(2^n)$	$f(10^n)$
1	.333333	.555555	.961905
2	.809524	.809524	.999600
3	.953488	.942029	.999996
4	.984674	.984674	.99999996
5	.993651	.996113	1.
⋮			
10	.999600	.999996	OVERFLOW
⋮			
20	.999975	1.	
⋮			
30	.999995	1.	
⋮			
100	.99999996	OVERFLOW	
⋮			
200	.999999998		
⋮			
1000	1		

probably this will happen before we can draw an appropriate conclusion about the value of the limit. Similar problems occur when numbers get too close to 0.

A second problem that can arise involves the effects of round-off errors. Thus, for instance, if we try to determine the limit of a quotient as x approaches infinity, the ratio may appear to approach or even reach a limit and then begin to diverge from it. This is illustrated in the next example.

EXAMPLE 2.10

Evaluate

$$\lim_{x \to \infty} \left(1 + \frac{1}{x} \right)^x$$

As in the previous example, we use several different sequences of x values to compare the results of each. In particular, we use $x = \{n^2\}$, $x = \{n^4\}$, $x = \{2^n\}$, and $x = \{5^n\}$.

If we examine each of the sequences of values in Table 2.8, we notice that each appears to be converging initially to a limit of approximately 2.718, but eventually each of the sequences begins to diverge. As a result, we must be extremely careful in selecting the apparent limit for any such function. Further, since most limits are not integers (as in this example), we cannot make any assumptions about the additional decimal places. In this particular case, the given limit turns out to be an irrational number, namely 2.718281828459045. . . .

TABLE 2.8

Values for $\lim_{x \to \infty}(1 + 1/x)^x$

n	$f(n^2)$	$f(n^4)$	$f(2^n)$	$f(5^n)$
1	2.	2.	2.25	2.488320
2	2.441406	2.637929	2.441406	2.665836
3	2.581175	2.701690	2.565785	2.707488
4	2.637929	2.712992	2.637929	2.716111
5	2.665836	2.716111	2.676990	2.717850
6	2.681464	2.717235	2.697345	2.718206
7	2.691053	2.717718	2.707739	2.718213
8	2.697345	2.717947	2.712991	2.718522
9	2.701690	2.718065	2.715632	2.712653
10	2.704813	2.718147	2.716955	2.728996
⋮				
15	2.712266	2.718224	2.718254	137.70407
⋮				
20	2.714893	2.718208	2.719074	OVERFLOW
⋮				
25	2.716111	2.718522	2.723218	
⋮				
30	2.716774	2.719424	3.363586	

SECTION 2.5 PROBLEMS

Use the program LIMFN from Section 2.4 or one of your modifications of it to investigate the behavior of each of the following limits to infinity.

1 $\lim\limits_{x \to \infty} \dfrac{2x^2 + 3x + 1}{5x^2 + x + 4}$

2 $\lim\limits_{x \to \infty} \dfrac{x^3 + x + 1}{3x^3 + 4}$

3 $\lim\limits_{x \to \infty} \dfrac{x^2}{2^x}$

4 $\lim\limits_{x \to \infty} \dfrac{2^x}{x^2}$

5 $\lim\limits_{x \to \infty} \dfrac{3 - 3^x}{5 - 5^x}$

6 $\lim\limits_{x \to \infty} \dfrac{3^x + 2x}{x^3 + 1}$

7 $\lim\limits_{x \to \infty} \dfrac{\sqrt{x^2 + 1}}{x}$

8 $\lim\limits_{x \to \infty} \dfrac{x - \cos x}{x}$

9 $\lim\limits_{x \to \infty} x \sin \dfrac{1}{x}$

10 $\lim\limits_{x \to \infty} \left(1 + \dfrac{1}{x}\right)^{5x}$

11 $\lim\limits_{x \to \infty} x^{1/x}$

12 $\lim\limits_{x \to \infty} \left(\dfrac{x^2}{x - 1} - \dfrac{x^2}{x + 1}\right)$

13 $\lim\limits_{x \to -\infty} \dfrac{2 - 1/x}{x^2 + 1}$

14 $\lim\limits_{x \to -\infty} \left(\dfrac{6}{\sqrt[3]{x}} + \dfrac{1}{\sqrt[5]{x}}\right)$

15 Consider $x^n/2^x$ for different values of n, say $n = 2, 3, 4, 10, \dots$. Use the program LIMFN and see if you can come to any conclusion about $\lim_{x \to \infty} x^n/2^x$ for any n. What about $\lim_{x \to \infty} x^n/3^x$? What about $\lim_{x \to \infty} x^n/a^x$

for any $a > 1$? What about $\lim_{x \to \infty} a^x/x^n$ for any $a > 1$? (Be careful with these — you may get OVERFLOW errors quickly.)

2.6
FINDING ROOTS BY THE BISECTION METHOD

One of the most important problems that arises throughout mathematics in general and calculus in particular is finding the roots of a function — that is, the values of x for which $f(x) = 0$. If the function is a quadratic polynomial, then the roots can be found easily using the quadratic formula. Similar, though considerably more complicated, formulas are available for finding the roots of any third- or fourth-degree polynomial. However, mathematicians have shown that no such formula *can* exist for polynomials whose degree is five or more. Therefore, if $f(x)$ is such a higher degree polynomial, then there is usually no way to find its roots exactly.

A similar problem occurs whenever $f(x)$ is a complicated expression such as a rational function, a function involving fractional powers, or a function involving trigonometric, logarithmic, or exponential expressions. Unfortunately, such cases arise frequently, and so a variety of techniques have been developed to find the roots of such functions. In the present section, we consider one such method, known as the Bisection Method. A different approach known as Newton's Method will be treated in a later section.

The Bisection Method is based on the fact that, in most cases, the graph of a function $f(x)$ crosses the x-axis at the point where the root occurs. Therefore, if the curve crosses the x-axis at the root $x = r$, then it must be above the x-axis, $f(x) > 0$, on one side of r and below the x-axis, $f(x) < 0$, on the other side of r. See Figure 2.4. The only exceptions to this corre-

FIGURE 2.4

The root of an equation

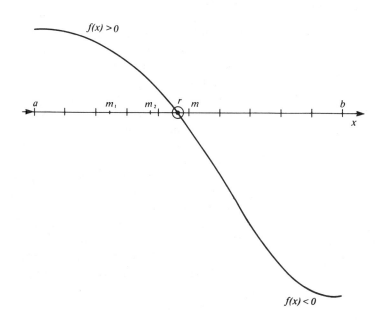

$f(x) > 0$

a m_1 m_2 r m b

x

$f(x) < 0$

FIGURE 2.5

The Bisection Method—
sequence of midpoints $m_1, m_2,$
$m_3, m_4, \ldots$ converging to root r

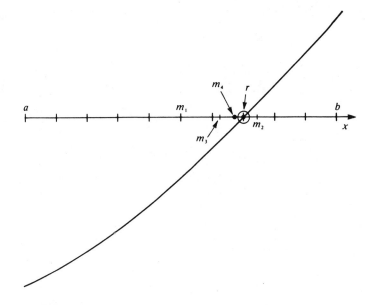

spond to the case where the curve is tangent to the x-axis at the root, and so the Bisection Method will not apply. See Figure 2.9.

Suppose we start with an interval $[a, b]$ such that $f(a)$ and $f(b)$ have opposite signs and f is continuous on the interval. Then there must be at least one point r within this interval, $x = r$, where the graph crosses the x-axis. We note that one way to express the fact that the signs of $f(a)$ and $f(b)$ are opposite is that $f(a) \cdot f(b) < 0$. The Bisection Method is based on the idea that if we now find the midpoint of the interval $[a, b]$, call it m, then the root must occur either in the left half-interval $[a, m]$ or in the right half-interval $[m, b]$. In Figure 2.4, the root is on the left. Having narrowed the interval containing the root by half, we then bisect the new half-interval at m_1, say, and so generate an interval one-quarter the size of the original one. If this procedure is continued indefinitely, then we produce a sequence of ever smaller subintervals, each containing the desired root. Eventually, the endpoints of the interval will agree to any desired degree of accuracy. Since the root is between the endpoints, its decimal representation must also have the same significant digits as the endpoints at that stage. This will be illustrated in the example below.

Before we proceed to the example, however, note that the key to determining which half of the interval we should use to continue the process is based on the observation earlier that if a root occurs in the interval $[a, b]$, then $f(a) \cdot f(b) \leq 0$. Thus, in Figure 2.5, it follows that $f(m_1) \cdot f(b) \leq 0$, so the right half-interval should be used. Then since $f(m_1) \cdot f(m_2) \leq 0$, the root must occur in the interval $[m_1, m_2]$. Further, since $f(m_3) \cdot f(m_2) \leq 0$, we would then focus on the subinterval $[m_3, m_2]$, and so on.

EXAMPLE 2.11   Suppose we seek a root of

$$x^{5/3} - x^{2/3} - 1 = 0.$$

To use the Bisection Method, we need to start with an interval where the function changes sign. We note that $f(0) = -1, f(1) = -1, f(2) = .59$, so that the initial interval should be $[a, b] = [1, 2]$. (We could have used $a = 0$ since the function changes sign on $[0, 2]$ also, but the larger the initial interval, the longer it will take for convergence to the root.) Table 2.9 provides the corresponding interval endpoints. See Figure 2.6.

TABLE 2.9

Bisection Method applied to
$f(x) = x^{5/3} - x^{2/3} - 1 = 0$

N	A	B
0	1	2
1	1.5	2
2	1.5	1.75
3	1.625	1.75
4	1.6875	1.75
5	1.6875	1.71875
6	1.6875	1.703125
7	1.695313	1.703125
8	1.6992195	1.703125
9	1.701172	1.703125
10	1.701172	1.702145
11	1.701172	1.701660
12	1.701416	1.701660
⋮		
15	1.701599	1.701660
⋮		
20	1.701607	1.701609

FIGURE 2.6

Graph of $f(x) = x^{5/3} - x^{2/3} - 1$

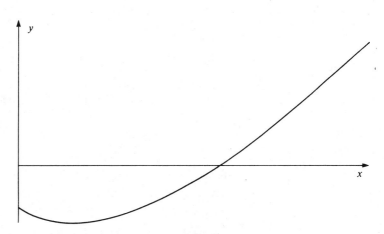

Note: x ranges from 0 to 3 in steps of .3
y ranges from -1.326 to 3.118 in steps of .4444

From these values, we conclude that the root lies between 1.701607 and 1.701609. Thus, correct to five decimal places, the root is $r = 1.70161$. Clearly, if greater accuracy were needed, then the process could be continued. However, it is worth noting that to get this degree of accuracy required 20 iterations, which is a fairly slow rate of convergence and is, in fact, typical of what we should expect whenever the Bisection Method is used.

One other point bears mentioning in connection with this example. If we examine the above table of values, we notice that the Bisection Method produces essentially two sequences of numbers, both of which converge to the desired root. One sequence converges from above and the other from below, so that the root r is always squeezed in between the corresponding terms of the two sequences for all values of n.

The following program BISECT, which was used to produce the table in Example 2.11, indicates how the Bisection Method can be easily implemented on the computer.

Program BISECT

```
10   REM CALCULATION OF ROOTS BY THE
     BISECTION METHOD
20   INPUT "THE INITIAL INTERVAL IS ";A,B
30   DEF FNY(X) = ...
40   PRINT "N","A","B")
50   LET M = (A + B)/2
60   IF FNY(A) * FNY(M) <= 0 THEN B = M
70   IF FNY(M) * FNY(B) <= 0 THEN A = M
80   PRINT N,A,B
90   GO TO 50
100  END
```

To use this program, you merely supply the initial interval across which the function changes sign and the desired function.

EXERCISE 9

Modify the program BISECT to provide an exit from the loop by testing whether the successive endpoints of the intervals agree to a predetermined number of significant digits (i.e., test whether the length of the interval is smaller than a certain value).

EXERCISE 10

Modify the program BISECT to test whether the given function actually changes sign across the interval. Also, provide a check to see if the root happens to occur at the midpoint m of any of the successive bisections. Finally, include the number of iterations as part of the output.

EXERCISE 11

If your computer allows the IF-THEN-ELSE statement, use it to replace lines 60 and 70 of the program BISECT with a single statement.

EXAMPLE 2.12   Find the first positive value of x for which $x \sin x = 2$.

Consider the function $f(x) = x \sin x - 2 = 0$. We find that $f(0) = -2 < 0$ and $f(7) = 2.5989 > 0$, so that there is a root in the interval $[0, 7]$. We thus apply program BISECT to find the results shown in Table 2.10. Since the desired root is between these last two values, we may conclude the correct value is approximately 6.59147. Alternatively, rather than rounding, we can also simply split the difference between the last two entries and use 6.5914695 as the answer. To verify that this is indeed close to the correct value for the root, we check by substituting back into the original function to find that $f(6.5914695) = .00001112$. The graph is shown in Figure 2.7.

We note that the search for an initial interval $[a, b]$ containing a root r of a function $f(x)$ can be accomplished most easily by just applying the program TABLE until a change of signs is detected.

There are several possible complications in using the Bisection Method. For one, if a function has several different roots in an interval I,

TABLE 2.10

Bisection Method applied to
$f(x) = x \sin x - 2 = 0$

N	A	B
1	3.5	7
2	5.25	7
3	6.125	7
4	6.5625	7
5	6.5625	6.78125
10	6.58984	6.59668
15	6.59134	6.59155
20	6.591466	6.591473

FIGURE 2.7

First positive root of $f(x) =$
$x \sin x - 2$

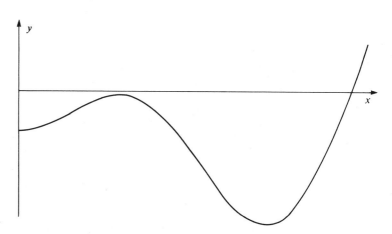

Note: x ranges from 0 to 7 in steps of .7
          y ranges from -6.814 to 2.402 in steps of .9217

then the Bisection Method will certainly converge to one of them, though not necessarily to the one wanted (say we want the largest or smallest root). Moreover, having located one such root, there is nothing in the method to indicate whether any other roots occur in that interval. That is something that has to be determined from an understanding of the graph of the function, and methods for doing so form an important part of calculus.

Furthermore, when a function has several roots that are extremely close together within an interval, testing that $f(a) \cdot f(b) \leq 0$ may not pick this up, as seen in Figure 2.8.

Finally, a root may occur at a point where the graph of the function is tangent to the x-axis, as shown in Figure 2.9 for the function $f(x) = x^2 - 4x + 4$. Clearly, there is no interval $[a, b]$ for which $f(a) \cdot f(b) \leq 0$, unless we happen to select either a or b as the root r. Therefore, the Bisection Method really does not apply in this situation. Fortunately, there are other methods available for locating roots of functions, some of which we will study in later chapters, that allow us to solve this type of problem.

FIGURE 2.8

Graph of a function having roots close to each other

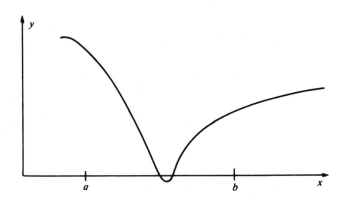

FIGURE 2.9

Point of tangency corresponding to a multiple root

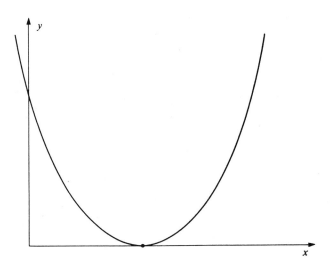

In applying any numerical method such as the Bisection Method to determine a root, it is important to realize that the best we can ever achieve is an approximation to the actual answer. At each repetition or *iteration* of the method, we obtain a better estimate of the value of the root. However, we cannot obtain an exact answer without performing an indefinite number of iterations, and that is not possible. Therefore, if we must content ourselves with an approximation to the true solution, then we should be able to estimate just how accurate the approximation is at each stage. This is not always possible to do, but with the Bisection Method, it is particularly simple.

Suppose we know that there is a root in the initial interval $[a, b]$. If we select the midpoint m of that interval, then it is obvious that the root r is no further than half of the length of the interval from m. That is, the error is at most $(b - a)/2$. For convenience, we call this first midpoint $m_1 = (a + b)/2$ and the corresponding error E_1. Thus, $E_1 \le (b - a)/2$. When we bisect the second time to obtain a new midpoint m_2 in either $[a, m_1]$ or in $[m_1, b]$, then the length of each subinterval is one-quarter of the original interval, and so the corresponding error $E_2 \le (b - a)/4$. On each successive iteration, the corresponding subinterval is half the length of the preceding one, and therefore it follows that the error diminishes by a factor of $1/2$ at each iteration. Thus, after n iterations of subdividing the interval, the approximation we obtain for the root r will have a maximum possible error of $E_n \le (b - a)/2^n$.

For example, if the original interval were $[a, b] = [0, 5]$, then after six repetitions of the method, the value for the midpoint m_6 would be off from the actual root by a maximum of $E_6 \le (b - a)/2^6 = 5/64 = .07812$ (which may not be very good accuracy). After ten iterations, the maximum error would be $E_{10} \le 5/2^{10} = 5/1024 = .00488$. Thus, after ten iterations, we would have an answer that is guaranteed to be within two significant decimal places, which is still not all that great. In fact, we saw in Example 2.11 that 20 iterations were required to achieve five decimal place accuracy.

In this analysis, we must realize that we are dealing with two interrelated sequences. The first, $\{m_n\}$, is a sequence of values converging to the root r. The second, $\{E_n\}$, is a sequence of error estimates that converges to 0 as $n \to \infty$. Incidentally, the error terms do not tell us exactly what the error is; they simply give us an estimate of the maximum possible size of the error.

We now turn around the above error analysis to estimate the number of iterations necessary to obtain a desired level of accuracy in the approximation. Suppose we want four-decimal accuracy. This means that the approximation must be within .00005 of the correct value. (We do not use .0001, as might be thought — we have to anticipate rounding in the fifth decimal place to obtain four correct decimals.) Therefore, we want to find the number of iterations n that guarantees that

$$E_n \le \frac{b - a}{2^n} \le .00005$$

or equivalently,

$$\frac{b-a}{.00005} = \frac{5}{.00005} \le 2^n$$

so that

$$2^n \ge 100,000$$

The first value of n for which this holds is $n = 17$, since $2^{17} = 131,072$. Therefore, 17 iterations will be required to guarantee us the desired degree of accuracy.

In general, we would have to solve for n from

$$E_n \le \frac{b-a}{2^n} \le E$$

where E is the acceptable error to find the number of iterations needed. If we have a selectively small initial interval $[a, b]$, then the number of iterations will be somewhat less.

However, if we want a high degree of accuracy, say 10 or 12 digits, then we still require a large number of iterations. Moreover, problems often arise that require finding a large number of different roots for a single function or for a set of different functions. In such cases, the Bisection Method is just too slow even with high speed computers available, and faster, more sophisticated methods are more appropriate. The Secant Method (see Problems 12 through 14 below) represents a slight improvement over the Bisection Method. Newton's Method in Section 3.4 converges much faster, but does not apply to all functions.

SECTION 2.6 PROBLEMS

Use the program BISECT or one of your modifications of it to locate the following roots correct to four decimal places.

1 The root of $x^3 - 3x + 1 = 0$ between 0 and 1

2 The root of $x^3 - 3x + 1 = 0$ between 1 and 2

3 The root of $x^3 - 3x + 1 = 0$ between -3 and 0

4 The largest root of $2x^3 - 4x^2 - 3x + 1 = 0$

5 The root of $x^4 - 5x^2 + 2x - 5 = 0$ between 2 and 3

6 The root of $x^5 + x^2 - 9x - 3 = 0$ between -2 and -1

7 All real roots of $x^4 - x - 2 = 0$

8 All real roots of $x^5 - 2x^2 + 4 = 0$

9 The root of $x \sin x - \cos x = 0$ between 0 and $\pi/2$

10 The smallest positive root of $x = \sqrt{x^3 - 5x + 4}$

11 The smallest positive root of $2x = \tan x$ between 0 and 1.5 ($< \pi/2$)

12 The Secant Method is based on the following construction: Consider the line drawn from $(a, f(a))$ to $(b, f(b))$ where $f(a)f(b) < 0$. The point m where this line crosses the x-axis is the next approximation to the root. Find the equation of this line, then set $y = 0$ and solve for the value of $x = m$ at this x-intercept.

13 Modify the program BISECT to use the value of m from the Secant Method instead of the midpoint from the Bisection Method.

14 Apply the Secant Method to Problems 1 through 11 and compare the number of iterations needed to converge to the root with the number needed using the Bisection Method.

2.7
LIMITS VIA THE DELTA-EPSILON APPROACH

In several of the preceding sections, we dealt with the concept of the limit of a function in an intuitive manner based on a sequential approach. This is fine as a means of picturing what a limit means. However, as was pointed out at the time, it lacks rigor in the sense that we cannot *prove* results conclusively using this method. Rather, we must adopt a more rigorous definition of limit that does provide the mathematical foundations for all of calculus.

Suppose we have a function $f(x)$ defined on an interval I except possibly at the point $x = a$. We say that the limit of $f(x)$ as x approaches a is L, written

$$\lim_{x \to a} f(x) = L$$

if, given any number $\varepsilon > 0$, there exists a corresponding number $\delta > 0$ such that:

$$|f(x) - L| < \varepsilon \quad \text{whenever} \quad 0 < |x - a| < \delta$$

Geometrically, this has the following interpretation: The expression $|f(x) - L|$ represents the vertical distance between the curve $y = f(x)$ and the horizontal line $y = L$. Thus, the inequality $|f(x) - L| < \varepsilon$ simply says that the distance between the curve and the line is less than the number ε given. This generates a horizontal band of height ε above and below the line $y = L$. If the limit of $f(x)$ is L as x approaches a, then the values of $f(x)$ should be close to L when x is near a, so that the curve must be within the band when x is close enough to a. But how close *is* close enough? This condition would only hold for values of x on either side of and sufficiently close to $x = a$. This in turn determines a vertical band (actually many different vertical bands are possible) with a certain width — we call any such appropriate width δ. The two bands together form a rectangle centered at the point (a, L) and, if the limit is indeed L, then the curve must lie entirely within the box — it cannot cross the top or bottom, only the sides of the box. See Figure 2.10.

FIGURE 2.10

The delta-epsilon definition of a limit

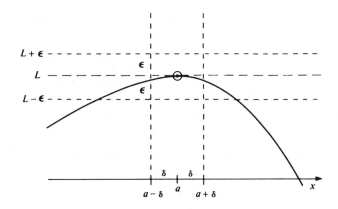

One of the problems most students have with this concept is that it is difficult to accept the notion that there is a delta for any given epsilon, provided that the limit exists. To make matters worse, it is almost impossible to find the delta value for a given epsilon for any but the simplest functions, and even then, the work is complicated enough to make one lose sight of what the problem is all about. Fortunately, we can use the computer to do the work for us. That is, we can use a program that will calculate (to any degree of accuracy desired) an appropriate value of δ corresponding to any given ε when the limit exists.

The idea behind the program is the following: Given a value for epsilon, call it E, we want to be sure that $|f(x) - L| < E$ for a certain range of values of x centered at $x = a$. In particular, we want to find out when $|f(x) - L| = E$ and, presumably, when x is taken any closer to a, $|f(x) - L| < E$. (There are cases where this is not true, and so the results are occasionally misleading.) To find this, we have to solve the two equations $f(x) - L = E$ and $f(x) - L = -E$. This can be done using the Bisection Method applied to the functions $g(x) = f(x) - L - E = 0$ and $h(x) = f(x) - L + E = 0$.

The resulting program is a fairly lengthy and complicated one. Therefore, we do not present it here. A prepackaged version under the name DELEPS should be available for your use. All that it requires is that you supply an interval $[x_1, x_2]$ for the function, the limit point a, the supposed limit L, the value of epsilon and the definition of the desired function. The program will respond with a value for delta (not necessarily the largest) or a message that no value was found that might suggest that the function does not have a limit or that the values for a and L do not match.

SECTION 2.7 PROBLEMS

Use the packaged program DELEPS (if available) to find a value of delta corresponding to the given values of epsilon for each function at the given point a.

1 $f(x) = \dfrac{1}{2}(3x - 1)$ with $a = 4$, $\varepsilon = .5, .02, .001$

2 $f(x) = x^2 + 4$ with $a = 1$, $\varepsilon = .1, .005$

3 $f(x) = (1 + x)^{1/x}$ with $a = 0$, $L = 2.71828$, $\varepsilon = .1, .01, .003$

4 $f(x) = \dfrac{x^2 - 25}{x + 5}$ with $a = -5$, $L = -10$, $\varepsilon = .05, .002$

5 $f(x) = \dfrac{8x^5 + 12x^4}{x^4}$ with $a = 0$, $L = 12$, $\varepsilon = .1, .005$

THREE
THE DERIVATIVE

3.1
THE DERIVATIVE VIA THE DEFINITION

In this chapter we turn our attention to the subject of the derivative of a function and the role that the computer can play in differentiation. First, though, it is important to keep in mind the fact that a computer's operations are totally numeric in nature, while the usual approach to derivatives in calculus is manipulative. Thus, we should not expect that if we give the computer a function, say $f(x) = x^3$, it will be able to respond with the derivative $f'(x) = 3x^2$. To do this effectively requires symbolic manipulations beyond the standard capabilities of BASIC and most other programming languages. Specialized software packages to perform "computer algebra" recently have become available using languages based on theories of artificial intelligence.

The best we can expect with BASIC is for the computer to calculate the *value* of the derivative, $f'(a)$, at any desired point $x = a$. To accomplish this, we resort to the definition of the derivative as

$$f'(x) = \lim_{h \to 0} \frac{f(x + h) - f(x)}{h}$$

We have seen that the computer can handle the limit of a function by calculating a sequence of values that converge to the limit L within a specific number of significant digits. Thus, to calculate the derivative of a function at a point $x = a$, we must have the computer calculate the value of the Newton difference-quotient $[f(a + h) - f(a)]/h$ for a sequence of values of h that approach 0.

As an example, suppose we use the function $f(x) = x^3$ at the point $x = a = 2$ with the sequence $h = 1/10^n$ for $n = 0, 1, 2 \ldots$. We thus obtain the set of values shown in Table 3.1. We see that this procedure produces a sequence of values that appears to be converging to the expected value, 12.

In theory, the computer should continue to calculate the Newton difference-quotient until two successive values have the identical digits — that is, until they agree to that number of decimal places. This approximate value would then be taken to be the value for $f'(x)$. We note that this calculation would only be accurate up to the number of digits computed. Thus, if the computer produces eight-digit numbers, we would only have the first eight digits of the exact value for $f'(x)$, so this approach produces only an

TABLE 3.1	H	QUOTIENT
Derivative of $f(x) = x^3$ at	1.	19.000000
x = 2 via the definition	.1	12.610000
	.01	12.060100
	.001	12.006001
	.0001	12.000600
	.00001	12.000060
	.000001	12.000006
	.0000001	12.000001

approximate value for $f'(x)$. For example, if for a given function the computer calculates $f'(a)$ to be 3.00000000, then there is absolutely no reason to conclude that $f'(a) = 3$ exactly. In fact, the next decimal place is just as likely to be a 3 or a 6 as it is to be another 0.

The key words in the last paragraph, however, are "in theory." In practice, unfortunately, the above procedure does not always produce a sequence of values that necessarily converges to $f'(a)$. The reason is that when h is close to 0, the difference $f(a + h) - f(a)$ is also very close to zero whenever the function is continuous. Consequently, the Newton quotient

$$\frac{f(a + h) - f(a)}{h}$$

is close to $0/0$. As a result, when the computer evaluates this ratio, the round-off errors in the numerator and denominator can be drastically magnified, and the sequence of values generated for the quotient may even diverge, rather than converge, to the value for $f'(a)$. This will occur if the process is carried too far and h is allowed to come too close to 0. Thus, if the above sequence were continued further, we would find that corresponding to $h = .0000001$, the quotient would come out to be 10 rather than 12 on certain computers. Further divergence would occur at each additional stage. The particular point at which this type of divergence may take place obviously is dependent on the computer as well as on the function, the point a, and the sequence of values for h.

The following program, DERIVF, is a simple way of implementing the above procedure to generate a table of values from which we hope to obtain the value of the derivative of a given function $f(x)$ at a desired point $x = a$.

Program DERIVF

```
10   REM DERIVATIVES USING THE DEFINITION
20   INPUT "WHAT IS THE LIMIT POINT? ";A
30   DEF FNY(X) = ...
40   PRINT "H","QUOTIENT"
50   FOR N = 1 TO 20
60   LET H = 1/2^N
70   LET Q = (FNY(A + H) - FNY(A))/H
80   PRINT H,Q
90   NEXT N
100  END
```

To use this program, we need to supply the desired point a at line 20 and to define the desired function FNY(X) at line 30. The program automatically uses the sequence H = $1/2^N$ at line 60.

For this type of procedure to work well, it is extremely important to select an appropriate sequence h that approaches 0 at a reasonably rapid rate. The use of $h = 1/2^n$ in the program probably is not the most effective choice — it requires 20 terms to reach a value of h smaller than .000001. However, its use certainly provides us with a feel for the convergence to the value of the derivative $f'(a)$. On the other hand, the use of a sequence that converges too rapidly produces a set of values that are almost certain to begin diverging long before any useful information can be deduced. For example, if we apply the program DERIVF to the function FNY(X) = X^3 at the point $a = 2$, using the sequence $h = 1/20^n$, then we obtain the results shown in Table 3.2. Without knowing that the derivative $f'(2)$ is supposed to be 12, there is no way that we could decide on exactly 12 as the correct value from this table. Thus, when the above approach is used, we must take great care in determining which value from the table to accept as the best approximation to the derivative. In particular, if the results begin to diverge, we have to use the value for $f'(x)$ that appears to be the most accurate and by doing so, we must realize that the choice is at best only a fair approximation to the true value.

Despite these difficulties, we need to keep in mind that the overwhelming majority of problems arising in the "real world" do not admit of closed form solutions that can be arrived at by formula. Instead, it is essential to resort to numerical techniques to obtain a solution, even if the solution may not be precise. In this regard, we note that there are many other numerical techniques available for approximating the values of derivatives that do not involve trying to evaluate a limit. However, since they are also approximations, an error of some degree is always inherent in the values so obtained. We will consider one such approach in the next section.

EXERCISE 1

Modify the program DERIVF so that the user can select the sequence h at line 60. Further, instead of automatically calculating a table with 20 entries, build in a loop that is completed once the value of h becomes smaller than some preselected test value, say .000001.

TABLE 3.2

Derivative of $f(x) = x^3$ at $x = 2$ using $h = 1/20^n$

H	QUOTIENT
.05	12.3025001
.0025	12.0150149
.0000125	12.0008886
6.25 E-6	11.9996071
3.125 E-7	12.0043754
1.5625 E-8	12.8746033
7.8125 E-10	12.9153444
3.9063 E-11	0

| EXERCISE 2 | Modify the above program DERIVF to compare two consecutive values of Q to see if they are within .00001 of each other. If they are, have the computer print an appropriate message about the approximate value of the derivative. |

| EXERCISE 3 | Modify the program you prepared in Exercise 2 to provide a means of checking for apparent divergence. That is, test whether successive differences Q1 − Q become smaller or larger in absolute value. |

| EXERCISE 4 | Modify the program DERIVF to provide a loop with appropriate INPUT statements on the point a and the acceptable tolerance E (.000001 was used for this in Exercise 1) so that you can have the computer evaluate the derivative of the function at a variety of points. |

| EXERCISE 5 | Modify the above programs so that the computer will generate a table of values for the derivatives at a range of points. |

SECTION 3.1 PROBLEMS

Use the program DERIVF or one of your modifications of it to estimate the value of the derivative of each of the following functions in Problems 1 through 10 at the indicated points. Wherever possible, compare the computer's answer with the actual one obtained by differentiation.

1 $f(x) = 2 - x^3$ at $x = 0, 1, 6$

2 $f(x) = 4x^2 - x + 6$ at $x = 2, 3$

3 $f(x) = \sqrt{x} + 1$ at $x = 0, 1, 9$

4 $f(x) = 1/x - 1$ at $x = 1, -1, 5$

5 $f(x) = 2/(x + 1)$ at $x = 1, -1.1$

6 $f(x) = x/(3x + 4)$ at $x = 0, 1$

7 $f(x) = (2x + 3)/(x + 4)$ at $x = -5, -3$

8 $f(x) = x + \sin x$ at $x = 0, \pi/2$

9 $f(x) = x \sin x$ at $x = 0, \pi/2$

10 $f(x) = (\sin x)^x$ at $x = \pi/4$

11 Use the results of Problem 10 to find the equation of the tangent line to the curve $y = (\sin x)^x$ at the point where $x = \pi/4$.

For each of the following functions, use DERIVF or one of your modifications of it to determine, if possible, whether the derivative exists at the indicated point.

12 $f(x) = x|x|$ at $x = 0$

13 $f(x) = \begin{cases} (1 + 1/x)^x & x \neq 0 \\ 2.71828\ldots & x = 0 \end{cases}$ at $x = 0$

14 $f(x) = \begin{cases} x^{\sin x} & x \neq 0 \\ 1 & x = 0 \end{cases}$ at $x = 0$

15 $f(x) = \begin{cases} \dfrac{\sin x}{x} & x \neq 0 \\ 1 & x = 0 \end{cases}$ at $x = 0$

16 $f(x) = \begin{cases} \dfrac{1 - \cos x}{x} & x \neq 0 \\ 0 & x = 0 \end{cases}$ at $x = 0$

3.2
DERIVATIVES USING INTERPOLATION

In the last section, the program DERIVF was used to calculate the value for the derivative of a function based on the definition of derivative. In most situations, this approach makes it possible to obtain a reasonable answer, though often the set of values calculated diverges, especially if h is allowed to approach too close to 0. The present section contains an alternative method for using the computer to calculate the derivative of a function. In most cases, it gives more accurate results in fewer steps and with less danger of divergence than a program based on the definition of derivative such as DERIVF.

The technique used for this method is known as *interpolation*. Before the advent of calculators, interpolation was commonly used to obtain values for trigonometric and logarithmic functions when the value for x was not specifically listed in the table. Interpolation is done by using two successive x values in the table to compute a value for an intermediate point.

For example, suppose we have the following entries from a table of values for the sine function (in radians):

x	$\sin x$
.17	.1692
.18	.1790

and we want to calculate the values for sin .175 and sin .173. In order to evaluate the first of these, we simply "split the difference" between the two table entries and use sin .175 = .1741. In order to evaluate sin .173, we calculate the value that would be $3/10$ of the way from .1692 to .1790 and so take sin .173 = .1721.

Essentially, what we do is to approximate the trig function by a linear function (line) through the two listed points and calculate the y value on the line corresponding to the intermediate x value. This procedure is called *linear interpolation*. It is also possible to extend the procedure to deal with polynomials of higher degree. Thus, just as two points uniquely determine the line that passes through them, three noncollinear points uniquely determine a quadratic polynomial, a parabola, that passes through them. Similarly, four points uniquely determine the cubic polynomial that passes through them, and so forth.

In each case, we can use the following interpretation. Suppose there is some function $f(x)$ that is quite complicated. We select a set of $n + 1$ points on the curve,

$$(x_0, f(x_0)), (x_1, f(x_1)), (x_2, f(x_2)), \ldots, (x_n, f(x_n))$$

and find the interpolating polynomial $P_n(x)$ of degree n that passes through these points. This polynomial will be an approximation to the original curve, just as the straight line was used as an approximation to the sine curve in the above example. That is, if we select any value of x other than the $n + 1$ values used to construct the interpolating polynomial, the value of $f(x)$ will be quite close to the value calculated by $P_n(x)$ using the polynomial. Details are given in the Appendix.

Without going into further details, we note that this interpolating polynomial can be used to approximate the value for the derivative of a given function. In particular, the function to be differentiated is approximated by a fourth-degree polynomial in the following program, and its derivative is calculated based on the derivative of the polynomial. The especially nice thing is that we do not even have to find the particular polynomial in order to use the program or the procedure. All we need to know is that the value of $f'(a)$ is approximately given by $[f(a - 2h) - 8f(a - h) + 8f(a + h) - f(a + 2h)]/12h$ or, in BASIC,

```
(FNY(A - 2 * H) - 8 * FNY(A - H)
 + 8 * FNY(A + H) - FNY(A + 2 * H))/(12 * H)
```

Program INTDRV

```
10   REM DERIVATIVES USING INTERPOLATION
20   INPUT "AT WHAT POINT A? ";A
30   DEF FNY(X) = ...
40   PRINT "H","APPROXIMATION"
50   FOR I = 1 TO 20
60   LET H = 1/2^I
70   LET R = (FNY(A - 2 * H) - 8
     * FNY(A - H) + 8 * FNY(A + H)
     - FNY(A + 2 * H))/(12 * H)
80   PRINT H,R
90   NEXT I
100  END
```

To use this program, simply supply the desired point a at line 20 and define the function you want, FNY(X), at line 30.

EXAMPLE 3.1

We apply this method to the function FNY(X) = X^3 with $a = 2$, as was done in the previous section. The results, in part, are found in Table 3.3.

TABLE 3.3	H	APPROXIMATION
Derivative of $f(x) = x^3$ at $x = 2$ using interpolation	.5	12
	.25	12
	.125	12.000001
	.0625	12
	.03125	11.9999998
	.015625	11.9999997
	.0078125	11.9999998
	.00195313	12.0000012
$I = 10$	.00097656	11.9999895
$I = 15$	3.05178 E-5	12.0004985
$I = 20$	9.536743 E-7	12.0065125
$I = 25$	2.980232 E-8	11.7708335
$I = 30$	9.313225 E-10	31.3333333

This example illustrates the great accuracy possible using the interpolation approach, though admittedly it is unreasonable to expect the exact answer to occur immediately on the first iteration, as it did here. It is more common to have the type of divergence portrayed in Table 3.3 as I increases, where the computed values gradually and then more quickly diverge away from the true result as the step approaches too close to zero.

We now consider a more typical example to show the use of this method and the associated program.

EXAMPLE 3.2   Find the derivative of

$$f(x) = \frac{\sqrt{x + 3}}{x^2 + 2} \quad \text{at } x = a = 1$$

When we apply the program INTDRV to the function $f(x)$, we obtain the results shown in Table 3.4. Based on these results, we see that the process clearly converges for the first half dozen or so terms and then gradually diverges. Table 3.4 therefore suggests that the approximate value for $f'(1)$ should be about $-.36111111$, correct to eight decimal places. By way of comparison, we note that the exact value calculated by formula is $-13/36 = -.3611111111\ldots$.

EXERCISE 6   Modify the program INTDRV the way you modified the program DERIVF in the previous section in Exercises 1 through 5.

TABLE 3.4	H	APPROXIMATION
Derivative of $f(x) = \sqrt{x + 3}/$	.5	-.360902394
$(x^2 + 2)$ at $x = 1$ using	.25	-.361278462
interpolation	.125	-.361125049
	.0625	-.361112042
	.03125	-.361111168
	.015625	-.361111112
	.0078125	-.361111115
	.0039063	-.361111135
	.00195313	-.361111086
	.00097656	-.361111106
	.00048828	-.361111642
$I = 15$	3.05176 E-5	-.361112595
	$\vdots$	$\vdots$
$I = 20$	9.53674 E-7	-.361083986
	$\vdots$	$\vdots$
$I = 25$	2.98023 E-8	-.368489587
	$\vdots$	$\vdots$
$I = 30$	9.31326 E-10	-.166666667

SECTION 3.2 PROBLEMS

Repeat Problems 1 through 10 of Section 3.1 using program INTDRV or one of your modifications of it. Again, compare the answers the program gives to the actual answers found by differentiation wherever possible.

3.3
HIGHER ORDER DERIVATIVES

In the present section, we consider the question of how the computer calculates the higher order derivatives for a given function $f(x)$ at a point $x = a$. We will examine two different procedures: one based on the definition of derivative, the other based on interpolation methods. As in the case of the first derivative, however, the interpolation approach is usually more effective.

We begin with the definition of the first derivative of $f(x)$:

$$f'(x) = \lim_{h \to 0} \frac{f(x + h) - f(x)}{h}$$

Since the second derivative is just the derivative of $f'(x)$, we have

$$f''(x) = \lim_{k \to 0} \frac{f'(x + k) - f'(x)}{k}$$

$$= \lim_{k \to 0} \left[\lim_{h \to 0} \frac{f(x + k + h) - f(x + k)}{h} - \lim_{h \to 0} \frac{f(x + h) - f(x)}{h} \right] \Big/ k$$

$$= \lim_{k \to 0} \lim_{h \to 0} \frac{f(x + k + h) - f(x + k) - f(x + h) + f(x)}{hk}$$

Since h and k are both approaching 0, it is convenient to set $k = h$ so that the above expression reduces to

$$f''(x) = \lim_{h \to 0} \frac{f(x + 2h) - 2f(x + h) + f(x)}{h^2}$$

In a totally similar manner, we can find

$$f'''(x) = \lim_{h \to 0} \frac{f(x + 3h) - 3f(x + 2h) + 3f(x + h) - f(x)}{h^3}$$

and

$$f^{\text{iv}}(x) = \lim_{h \to 0} \frac{f(x + 4h) - 4f(x + 3h) + 6f(x + 2h) - 4f(x + h) + f(x)}{h^4}$$

EXERCISE 7   Derive the above expression for $f'''(x)$.

If we examine the above expressions carefully, several patterns become clear. First, the function is being evaluated at the successive points $x, x + h, x + 2h, x + 3h, \ldots, x + nh$ for the nth derivative $f^{(n)}(x)$. Second, the signs of the terms strictly alternate, always starting with a positive term. Third, the denominator is precisely h^n for the nth derivative term. Fourth, the coefficients of the expressions can be arranged in the following pattern:

$$
\begin{array}{ccccccc}
 & 1 & & 1 & & & n = 1 \\
 & 1 & & 2 & & 1 & n = 2 \\
1 & & 3 & & 3 & & 1 \quad n = 3 \\
1 & & 4 & & 6 & & 4 \quad 1 \quad n = 4
\end{array}
$$

These numbers are known as the binomial coefficients since they arise as the coefficients in $(a + b)^n$ and, when arranged in the triangular pattern shown above, form Pascal's Triangle. The key to these numbers is that, in Pascal's Triangle, every interior entry is precisely the sum of the two numbers above it. Thus, $6 = 3 + 3$, $4 = 1 + 3$, and so on. The next line would then consist of the entries: 1   5   10   10   5   1, and the corresponding formula for the fifth derivative would then be

$$f^{\text{v}}(x) = \lim_{h \to 0} \frac{f(x + 5h) - 5f(x + 4h) + 10f(x + 3h) - 10f(x + 2h)}{h^5}$$
$$+ \frac{5f(x + h) - f(x)}{h^5}$$

EXERCISE 8   What is the formula for the sixth derivative?

The trouble with using any of the above formulas is that as $h \to 0$, the corresponding quotient approaches $0/0$ at a much faster rate than it did in calculating values for the first derivative. As a result, the sequence generated is going to diverge much sooner and therefore will provide less useful information. In fact, the higher the order of the derivative, the faster the divergence appears. Consequently, while interesting, the above approach is often not especially useful. However, for those who wish to experiment with it, we have a program to calculate the first few higher order derivatives.

Program DERIVN1

```
10    REM HIGHER ORDER DERIVATIVES VIA THE
      DEFINITION
20    INPUT "WHAT IS THE ORDER N? ";N
30    INPUT "WHAT IS THE POINT A? ";A
40    DEF FNY(X) = ...
50    PRINT "H","X","QUOTIENT"
60    FOR I = 1 TO 20
70    LET H = 1/2^I
80    LET Q(2) = (FNY(A + 2 * H)
      - 2 * FNY(A + H) + FNY(A))/H^2
90    LET Q(3) = (FNY(A + 3 * H)
      - 3 * FNY(A + 2 * H) + 3 * FNY(A + H)
      - FNY(A))/H^3
120   PRINT H,A + H,Q(N)
130   NEXT I
140   END
```

To use this program, select the value for the order of the derivative n (either 2 or 3) and the point where the derivative is to be calculated, a. Finally, define the desired function at line 40.

EXERCISE 9

Modify the above program to include the fourth and fifth derivatives. Also, include a check that n does not exceed the maximum allowed.

EXAMPLE 3.3

We again consider the function $f(x) = x^3$ at the point $a = 2$. For comparison, we know that the second derivative should turn out to be $f''(2) = 12$. Using the above program, we obtain, in part, for the second derivative the results shown in Table 3.5. These values certainly appear to be converging to the expected value of 12. Eventually, though, as h gets too close to zero, the values begin to diverge. As a result, it is difficult to actually predict that the correct value for the second derivative should be 12 if we did not know this from other sources.

As mentioned above, we usually will be more successful in calculating the values for the higher derivatives if we make use of interpolation methods instead of the definition of derivative. To do this efficiently involves the use of a higher degree interpolating polynomial, namely one of sixth degree based on seven points. As in the previous section, we fortunately do not have to actually determine this polynomial to make use of the method. The appropriate program is given below.

TABLE 3.5

Second derivative of $f(x) = x^3$
at $x = 2$ using the limit
definition

H	X	QUOTIENT
.5	2.5	15.
.25	2.25	13.5
.125	2.125	12.75
.0625	2.0625	12.375
.03125	2.03125	12.1875115
.015625	2.015625	12.09375
.0078125	2.0078125	12.0469361
.0039063	2.0039063	12.025466
.0019531	2.0019531	12.0117188
.0009766	2.0009766	12.0312501
.0004883	2.0004883	12.0625001
.0002441	2.0002441	11.8750001

Program DERIVN2

```
10    REM HIGHER ORDER DERIVATIVES VIA
      INTERPOLATION
20    INPUT "WHAT IS THE ORDER N? ";N
30    INPUT "WHAT IS THE POINT A? ";A
40    DEF FNY(X) = ...
50    PRINT "H","X","QUOTIENT"
60    FOR I = 1 TO 20
70    LET H = 1/2^I
80    LET Q(1) = -(FNY(A - 3 * H) - 9
      * FNY(A - 2 * H) + 45 * FNY(A - H)
      - 45 * FNY(A + H) + 9 * FNY(A + 2 * H)
      - FNY(A + 3 * H))/(60 * H)
90    LET Q(2) = (FNY(A - 3 * H) - 13.5
      * FNY(A - 2 * H) + 135 * FNY(A - H)
      - 245 * FNY(A) + 135 * FNY(A + H)
      - 13.5 * FNY(A + 2 * H) + FNY(A + 3
      * H))/(90 * H^2)
100   LET Q(3) = (FNY(A - 3 * H) - 8
      * FNY(A - 2 * H) + 13 * FNY(A - H)
      - 13 * FNY(A + H) + 8 * FNY(A + 2
      * H) - FNY(A + 3 * H))/(8 * H^3)
110   LET Q(4) = (-FNY(A - 3 * H) + 12
      * FNY(A - 2 * H) - 39 * FNY(A - H)
      + 56 * FNY(A) - 39 * FNY(A + H)
      + 12 * FNY(A + 2 * H) - FNY(A + 3
      * H))/(6 * H^4)
120   LET Q(5) = -(FNY(A - 3 * H) - 4 * FNY(A
      - 2 * H) + 5 * FNY(A - H) - 5 * FNY(A
      + H) + 4 * FNY(A + 2 * H) - FNY(A + 3
      * H))/(2 * H^5)
```

```
130   LET Q(6) = (FNY(A - 3 * H) - 6 * FNY(A
      - 2 * H) + 15 * FNY(A - H) - 20
      * FNY(A) + 15 * FNY(A + H) - 6
      * FNY(A + 2 * H) + FNY(A + 3 * H))/H^6
140   PRINT H,A + H,Q(N)
150   NEXT I
160   END
```

As in the preceding programs involving computation of derivatives, the calculations performed in this program also involve quotients that approach the form 0/0 rather rapidly. As a consequence, the sequence of numbers generated initially converge to the correct value and then diverge. We may therefore be faced with having to select among the printed values for the one that is closest to $f^{(n)}(a)$, rather than being presented with the correct answer.

EXAMPLE 3.4   We consider the function $f(x) = x^3$ at $a = 2$ as an example where we apply the above program to approximate the second derivative. The results are shown, in part, in Table 3.6. When we compare these results to the ones in Example 3.3 that use the definition of the second derivative as a limit, we see that the interpolation approach is far more accurate and more resistant to divergence as h approaches close to 0. Eventually, though, the sequence here also begins to diverge. However, before it does, we certainly have more useful information upon which to draw a conclusion about the value of the second derivative of the function. A much better challenge is presented in the next example.

EXAMPLE 3.5   Find an approximate value for the third derivative of

$$f(x) = \sqrt{\sin x + \tan x} \quad \text{at } x = a = 1$$

TABLE 3.6
Second derivative of $f(x) = x^3$
at $x = 2$ using interpolation

H	X	QUOTIENT
.5	2.5	12.
.25	2.25	11.999999
.125	2.125	12.000001
.0625	2.0625	11.999997
.03125	2.03125	11.999995
.015625	2.015625	12.000003
.0078125	2.0078125	11.999915
.0039063	2.0039063	11.998223
.001953	2.001953	11.993837
.000976	2.000976	11.980035

To solve this problem in closed form by hand is extremely difficult due to the nature of the function and its derivatives. Therefore, we might be tempted to apply the program DERIVN2. If we do, however, the computer will respond with an error message, possibly indicating an attempt to take the square root of a negative number. The reason for this is that while $f(x)$ is defined at $a = 1$, the formulas being used to approximate the derivatives involve calculating $f(x)$ at $a - 3h, a - 2h, a - h, \ldots, a + 3h$. Clearly, at least one of these points is such that the expression inside the radical is negative.

This difficulty can be avoided by starting closer to $a = 1$ than happens with a step of $h = .5$. Thus, if we change line 60 to read

```
60   FOR I = 3 TO 20
```

the program will function well and produces the results shown in Table 3.7. These results suggest the $f'''(1) = -12.886$, correct to three decimal places.

Incidentally, in using this program, DERIVN2, or the preceding one, DERIVN1, care should be taken to use only functions that have the appropriate number of derivatives. If we ask the computer to calculate a derivative that does not exist, the results will be at best highly misleading. Thus, for example, we should not attempt to find the third or higher derivative of the function $f(x) = x^{5/2}$ at the point $x = a = 0$.

EXERCISE 10

Modify the above program DERIVN2 so that it comes out of a loop with an answer when two successive values of $Q(n)$ agree to a preselected number of digits.

EXERCISE 11

Modify the program DERIVN2 so that it comes out of a loop as soon as successive sets of values of $Q(n)$ begin to diverge.

EXERCISE 12

Modify the program DERIVN2 to include an extra loop so that it will calculate the value of each derivative for $n = 1, 2, \ldots .6$.

TABLE 3.7

Third derivative of $f(x) =$ $\sqrt{\sin x + \tan x}$ at $x = 1$ using interpolation

H	A + H	QUOTIENT
.125	1.125	-10.912527
.0625	1.0625	-12.815519
.03125	1.03125	-12.882938
.015625	1.015625	-12.886520
.0078125	1.0078125	-12.886719
.0039063	1.0039063	-12.881836
.0019531	1.0019531	-12.67187
.000976	1.000976	-12.75
.000488	1.000488	-15.5

SECTION 3.3 PROBLEMS

Use either program DERIVN1 or DERIVN2 or one of your modifications of either to attempt to approximate the values of the following higher order derivatives at the indicated points.

1 $f(x) = 15x^4 - 4x^2 + x$, f'' at $x = 0, 5$

2 $f(x) = 8x^2 - \pi^2$, f'' at $x = 0, 2$

3 $f(x) = 3x^8 - 2x^5$, f'' at $x = 0, 3, 10$

4 $f(x) = 3x^8 - 2x^5$, f''' at $x = 0, 3$

5 $f(x) = 2x^5 + 4x^3 - 6x^2$, f''' at $x = 1, 2, 3$

6 $f(x) = x^3 + 8x^2 - \dfrac{2}{x^4}$, f'' at $x = 1, -1$

7 $f(x) = \sqrt[5]{10x + 7}$, f''' at $x = 0$

8 $f(x) = x^{3/2} - 2x^{1/2} + 4x^{-1/2}$, f'' at $x = 1, 5.973$

9 $f(x) = \sin(x^2 + \sqrt{x})/\sqrt{\sin x}$, f'' at $x = 1, \pi/2$

10 $f(x) = \tan(x/2)$, f''' at $x = 0, \pi/6$

11 $f(x) = x^{\sin x}$, f'' at $x = \pi/2$

12 $f(x) = x^{\sin x}$, f''' at $x = \pi/2$

3.4
FINDING ROOTS BY NEWTON'S METHOD

In the last chapter, we discussed the problem of determining the roots for a function $f(x)$ using the Bisection Method. At the time, we saw that while the method is an easy one to apply, the rate of convergence to the root is relatively slow.

An alternate approach to determining the roots for a function is known as *Newton's Method*. Geometrically, it is based on the following procedure. Suppose x_0 is an initial estimate of the value of the root r. This determines a point $P(x_0, f(x_0))$ on the graph of the function, as shown in Figure 3.1. Using calculus, we know that the tangent line to the curve at P has slope $f'(x_0)$, and therefore the tangent line drawn to the curve at this point P is given by

$$y - f(x_0) = m(x - x_0) = f'(x_0)(x - x_0)$$

The situation shown in the figure is quite typical in that the point where the tangent line crosses the x-axis is much closer to the root r than the original estimate x_0 is. As a result, it makes sense to use the value for this x-intercept as a new and better estimate for the root. By way of notation, we denote the intercept by $Q(x_1, 0)$, so that the equation of the tangent line yields

$$0 - f(x_0) = f'(x_0)(x_1 - x_0)$$

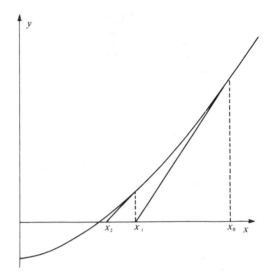

In turn,

$$x_1 - x_0 = \frac{-f(x_0)}{f'(x_0)}$$

assuming that $f'(x_0) \neq 0$, and therefore,

$$x_1 = x_0 - \frac{f(x_0)}{f'(x_0)}$$

This new value x_1 will then be used to determine a new point on the curve that in turn leads to a new tangent line whose x-intercept x_2 is an even better approximation to r, and so on, as shown in Figure 3.1.

As we will see, when Newton's Method works, the rate of convergence is exceptionally rapid and, in fact, eventually almost doubles the number of significant digits at each successive iteration.

EXAMPLE 3.6

We again consider the equation $f(x) = x^{5/3} - x^{2/3} - 1 = 0$ that we studied in the last chapter, where we used the Bisection Method to determine the root near 1.70161. Again, since $f(0) = -1$, $f(1) = -1$, and $f(2) = .59$, we might be tempted to use $x_0 = 0$, $x_0 = 1$, or $x_0 = 2$ as the initial estimate for r. To get a feel for the effectiveness of Newton's Method, we will try it using several of these possible initial estimates as well as $x_0 = 10$ and $x_0 = 100$, which are poor initial guesses indeed. The different sets of results are shown in Table 3.8. We note that in each case, the sequence of values obtained quickly converges to the root at 1.701607. Further, even for an exceptionally poor initial approximation, the rate of convergence is dramatic. This is especially striking when we compare these results with the fact that it took a full 20 iterations to obtain the same answer using the Bisection Method.

TABLE 3.8

Convergence of Newton's
Method for $f(x) =$
$x^{5/3} - x^{2/3} - 1 = 0$

	X0 = 1	X0 = 2	X0 = 10	X0 = 100
$n = 1$	2.	1.722470	4.509652	40.389407
$n = 2$	1.722479	1.701729	2.440109	16.570843
$n = 3$	1.701729	1.701607	1.802573	7.091837
$n = 4$	1.701607	1.701607	1.704321	3.390516
$n = 5$	1.701607	$\vdots$	1.701609	2.065771
$n = 6$	$\vdots$		1.701607	1.731491
$n = 7$			1.701607	1.701857
$n = 8$			$\vdots$	1.701607
				1.701607
				$\vdots$

It is worth noting that we could not have used $x_0 = 0$ as the initial approximation in the example since $f'(x_0) = f'(0)$ is not defined. Moreover, the choice of a negative value for x_0 in *this* example would create problems because of the way that the computer evaluates exponents. To illustrate this type of difficulty, consider $\sqrt[3]{-8} = (-8)^{1/3}$. We know that this has a value of -2. However, if we tried to do this on a computer or on most handheld calculators, we would get an error message. The algorithm used to evaluate exponents involves using the logarithm of the base, and since logs are only defined for positive arguments, $\log(-8)$ is meaningless.

The following program, NEWTON, is a simple way of implementing Newton's Method on the computer.

Program NEWTON

```
10   REM NEWTON'S METHOD FOR FINDING ROOTS
20   INPUT "THE INITIAL VALUE IS ";X0
30   DEF FNY(X) = ...
40   DEF FND(X) = ...
50   X1 = X0 - FNY(X0)/FND(X0)
60   PRINT X1
70   LET X0 = X1
80   GO TO 50
90   END
```

To use this program, it is necessary to supply the computer with both the desired function FNY(X) and its derivative FND(X) at lines 30 and 40, respectively, as well as the initial estimate X0 of the root.

The primary difficulty in using Newton's Method arises because of the presence of $f'(x_0)$ in the denominator of the formula. Certainly, if $f'(x_0) = 0$ at any stage in the sequence generated by the method, the formula would not be defined. However, severe problems can occur if $f'(x)$ is just relatively close to 0, even if it is never equal to 0. The reason for this can best be seen geometrically. If $f'(x) \approx 0$, then the resulting tangent line is close to being horizontal. Therefore, the corresponding x-intercept will be at a considerable distance from the previous value. Several different things can therefore happen. If the function is basically monotonic, then the sequence of values will still likely converge to a root. The values will simply return to the neighborhood of that root. Sometimes, though, the sequence of values will simply diverge instead. On the other hand, if the function is very "wiggly," such as a sine curve, then Newton's Method *may* converge to a root, though conceivably to one far removed from the starting point even if the initial estimate is close to some other root. See Figure 3.2.

Further, examples have been constructed where the process does not converge at all — the x-intercepts may literally bounce back and forth forever. Such a case occurs with the function $f(x) = x^3 - 4x = x(x^2 - 4)$, which has roots at $x = 0, 2, -2$. However, should we happen to apply Newton's Method starting with $x_0 = .8944270610809326$, then we would obtain $x_1 = -.8944270610809326$, from which $x_2 = x_0$, and so on indefinitely. See Figure 3.3. Admittedly, to have this happen, we either have to be incredibly unlucky or have worked hard to achieve it. Nevertheless, it is something that can happen.

EXERCISE 13

Modify the above program NEWTON to provide an exit from the loop when two successive values x_0 and x_1 are sufficiently close.

EXERCISE 14

Modify the program NEWTON to provide a check on whether the derivative FND(X0) is zero. If the derivative is zero, try to fix things by giving the last value of x calculated, x_0, a little nudge (that is, let $x_0 = x_0 + .00001$, say) in the hopes that the zero derivative can be avoided.

EXERCISE 15

Modify the program NEWTON to terminate the loop after a predetermined number of iterations have taken place that would suggest that the method is not converging.

FIGURE 3.2

Newton's Method when $f'(x_0) \approx 0$

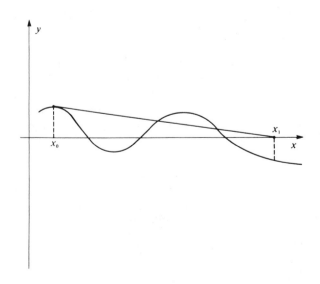

FIGURE 3.3

Newton's Method "stuck" in a loop for $f(x) = x^3 - 4x$

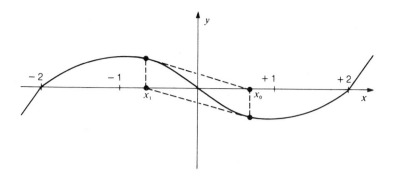

Several other points bear mentioning regarding the selection of the initial estimate for Newton's Method. The example above might be somewhat misleading since the choice of $x_0 = 100$ still produced the correct root in a relatively small number of iterations. It is possible, however, that for certain functions, if x_0 is chosen too far from r, then the method will produce a sequence that does not converge.

Furthermore, in the last chapter, we suggested that the program TABLE should be used in conjunction with the program BISECT to find an initial interval containing a root of the function under study. The same idea makes sense when using Newton's Method. Program TABLE can be applied (or the modification of it requested in Exercise 4 of Section 2.1) to bracket a root of $f(x)$. Since Newton's Method requires only a single initial estimate, we can simply locate such an interval and select its midpoint as the choice for x_0 in program NEWTON.

EXERCISE 16

Combine the programs TABLE and NEWTON into a single program that locates an interval containing a root, selects the midpoint as the choice for x_0, and then applies Newton's Method to zero in on the root. Remember to provide an exit from the program if no such interval can be found using the value of h given.

The program NEWTON can also be used effectively in conjunction with the program INCDEC from Section 2.2 to zero in on the roots of the derivative, $f'(x) = 0$, once INCDEC has provided the initial estimates for the location of critical points. If anything, NEWTON would be a better choice than BISECT in such a situation, in part because of the speed of convergence. More importantly, the output for INCDEC is just a single value near the root (which is all that NEWTON requires for x_0) while BISECT requires an initial interval containing the root. When this approach is employed, however, remember that we are seeking the roots of the derivative. Therefore, the functions supplied to the computer for the program NEWTON should be $f'(x)$ as FNY(X) and $f''(x)$ as FND(X).

EXAMPLE 3.7	Find the smallest positive angle (in radians) for which $g(x) = x \sin x$ attains the value of 2.

We consider the function $f(x) = x \sin x - 2$ and seek a root of it. We first apply the program TABLE to this function on the interval $[-2, 18]$ with step $h = .5$ and so find that

$f(x)$ changes sign between 6.5 and 7

$f(x)$ changes sign between 9 and 9.5

$f(x)$ changes sign between 12.5 and 13

$f(x)$ changes sign between 15.5 and 16

Based on this, we conclude that the smallest positive root should lie between 6.5 and 7. Notice that we said "should" rather than "must." It is conceivable that the function has two roots close together (or even a multiple root) within the space of a single step of length $h = .5$. If this were the case, the double sign change would not be picked up using a step of $h = .5$. The use of a smaller step size, say $h = .1$ or $h = .01$, would be more likely to detect such a case, but would still not absolutely guarantee it. It would, though, take much longer to run the program and would generate far more data to be analyzed. In fact, it is theoretically impossible to cover every possible eventuality using a finite step h. In practice, we simply try to compromise between using a reasonably small step h that is handled in a reasonably short length of time on the computer and the slight possibility of missing a root.

Since the smallest positive root is probably between 6.5 and 7, we take the midpoint of this interval as $x_0 = 6.75$. Program NEWTON then generates the following results:

```
6.589793
6.591468
6.59146781
```

The last value is correct to 8 significant decimal places.

FIGURE 3.4

Graph of $f(x) = x \sin x$

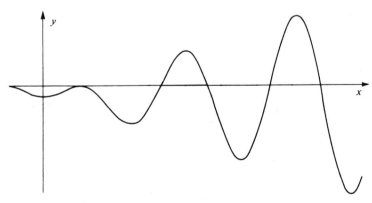

Note: x ranges from -2 to 18 in steps of 2
y ranges from -19.28 to 12.149 in steps of 3.1429

We mentioned earlier that a variety of problems can arise when Newton's Method is applied to a sinusoidal function, and the one in this example certainly falls into this category as shown in Figure 3.4. Just to illustrate what can happen, Table 3.9 lists the outputs obtained from NEWTON for a variety of other possible choices for x_0. These results should be interpreted in terms of the graph of the function shown in Figure 3.4. Admittedly, the cases in Table 3.9 are fairly extreme, but they do indicate what can happen if the initial choice for x_0 or some subsequent iterate is at a point where the curve is too flat and $f'(x)$ is too close to zero.

TABLE 3.9

Convergence of Newton's Method for $f(x) = x \sin x - 2 = 0$

X0 = 6	X0 = 5.5	X0 = 5.4	X0 = 5.3
6.6707	7.34217	7.72539	8.34429
6.59124	6.35703	4.86949	10.1037
6.59147	6.59574	-25.2849	9.12093
6.59147	6.59147	-25.2120	9.20767
	6.59147	-25.2122	9.20578
		-25.2122	9.20578

X0 = 4	X0 = 3	X0 = 1.5
2.50886	2.44266	1.95647
2.14821	2.09339	2.93905
1.55055	1.05032	2.41296
1.98671	1.83366	2.06624
3.61681	2.30245	.28975
3.61681	2.30245	.28975
⋮	⋮	⋮
-12.7242	-53.3696	-9.20578
(19 iterations to converge)	(23 iterations to converge)	(54 iterations to converge)

SECTION 3.4 PROBLEMS

1–10 Apply program NEWTON or one of your modifications of it to Problems 1 through 10 of Section 2.6 on the Bisection Method. In each case, compare the number of iterations needed to obtain the equivalent degree of accuracy.

11 Apply Newton's Method to find the smallest positive solution of $\sqrt{x} = \cos x$. (Hint: Derivative of $\cos x$ is equal to $-\sin x$.)

12 Apply Newton's Method to find the smallest positive solution of $x = \sqrt{\cos x}$.

13 Apply Newton's Method to find the smallest positive solution of $x = \cos \sqrt{x}$.

3.5
THE MEAN VALUE THEOREM

According to the Mean Value Theorem,

If a function $f(x)$ is continuous on the closed interval $[a, b]$ and differentiable on the open interval (a, b), then there exists a point c in (a, b) such that

$$f'(c) = \frac{f(b) - f(a)}{b - a}$$

The Mean Value Theorem only guarantees the existence of the number c; it does not provide any further clue to its location beyond the fact that it must lie somewhere in the open interval (a, b). In fact, for all but the simplest functions, there are no straightforward algebraic methods for explicitly finding c.

The program MVT is designed to determine the value of c via approximation methods. In order to use it, you must supply the computer with the desired function FNY(X), with its derivative FND(X), and with its second derivative FNS(X) using DEF statements at lines 20, 30, and 40, respectively. In addition, you have to provide the endpoints of the interval, a and b. The computer will respond with the calculated value for c.

Program MVT

```
10   REM FINDING C FROM THE MEAN VALUE
     THEOREM
20   DEF FNY(X) = ...
30   DEF FND(X) = ...
40   DEF FNS(X) = ...
50   INPUT "WHAT IS THE INTERVAL A,B? ";A,B
60   DEF FNB(X) = FND(X) - (FNY(B) - FNY(A))/
     (B - A)
70   LET H = B - A
80   FOR X = A TO B - H STEP H
90   IF FNB(X) = 0 THEN 190
100   IF FNB(X) * FNB(X + H) < 0 THEN 140
110   NEXT X
120   LET H = H/10
130   GO TO 80
140   LET X = X + H/2
150   LET C = X - FNB(X)/FNS(X)
160   IF ABS(C - X) < .0001 THEN 200
170   LET X = C
180   GO TO 160
190   LET C = X
200   PRINT "THE VALUE OF C IS ";C
210   END
```

We note that the above program is an application of the program NEWTON used in the previous section to locate roots using Newton's

Method. In the present case, we are looking for the particular value of x, namely c, which makes

$$f'(x) - \frac{f(b) - f(a)}{b - a} = 0$$

In order to accomplish this, the program constructs a new function

```
FNB(X) = FND(X) - (FNY(B) - FNY(A))/(B - A)
```

at line 60 and applies Newton's Method to it to find its root. Lines 70–110 are designed to locate an appropriate subinterval of (a, b), across which the newly constructed function FNB(X) changes sign so that Newton's Method can be applied.

As an example, suppose we select the function

$$f(x) = (x^3 + 1)^2$$

on the interval $[5, 11]$. See Figure 3.5. In order to use the program MVT, we have to supply

```
20   DEF FNY(X) = (X^3 + 1)^2
30   DEF FND(X) = 2 * (X^3 + 1) * 3 * X^2
40   DEF FNS(X) = 30 * X^4 + 12 * X
```

After we supply the values for a and b, namely 5 and 11, the computer will respond with

```
THE VALUE OF C IS 8.66218
```

FIGURE 3.5

Mean Value Theorem applied to
$f(x) = (x^3 + 1)^2$ **on** $[5, 11]$

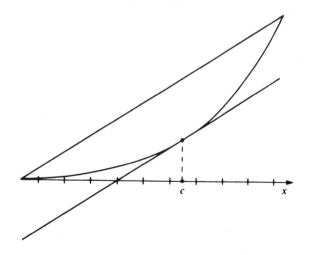

Note: x ranges from 5 to 11 in steps of .6
y ranges from $- .649492$ to $.177422E\ 7$ in steps of .127654
$c = 8.66218$

EXERCISE 17

Modify the program MVT to check for possible division by 0 at line 150. If FNS(X) is zero, supply either an exit with an appropriate message or make a minor change in the value of FNS(X) and allow the program to continue.

EXERCISE 18

The instructions in lines 70–110 are intended to find an interval across which FNB(X) changes sign. It is possible to use a function for FNY(X) for which FNB(X) is strictly positive or strictly negative between a and b. (A trivial example is FNY(X) = 10 $*$ X on [2, 5]). In such a case, this loop in the program may be traversed indefinitely. Modify the program to exit from the loop after a reasonable number of iterations.

EXERCISE 19

The instructions in lines 140–180 involve the computations using Newton's Method. It sometimes happens that the computer will get into an infinite loop at such a stage, either through round-off errors or because the numbers involved are exceptionally close to 0. Modify the program to prevent such an infinite loop from occurring while simultaneously allowing a reasonable number of iterations to be carried out.

EXERCISE 20

The program MVT provides for the calculation of just one value for c, although it is possible for a particular function to satisfy the Mean Value Theorem at several distinct points. Modify the program to permit it to search for additional values of c after the first has been found.

SECTION 3.5 PROBLEMS

Apply the program MVT or one of your modifications of it to each of the following functions with the indicated intervals.

1 $f(x) = x^3 + 1$ on $[-2, 4]$
2 $f(x) = 5x^2 - 3x + 1$ on $[1, 3]$
3 $f(x) = x + 4/x$ on $[1, 4]$
4 $f(x) = -x^2 + 8x - 6$ on $[2, 3]$
5 $f(x) = x^{2/3}$ on $[1, 8]$
6 $f(x) = 1/(x - 1)^2$ on $[2, 5]$
7 $f(x) = \sqrt{4x + 1}$ on $[2, 6]$
8 $f(x) = x^{1/3} - x^{3/5}$ on $[1, 8]$
9 $f(x) = x^6 + 12x^4 + (x + 1)/(x^3 - 2)$ on $[3, 5]$
10 $f(x) = x \sin x$ on $[0, \pi]$ (Hint: The derivative of $\sin x$ is $\cos x$.)

OPTIMIZATION OF FUNCTIONS

3.6

One of the most important applications of the derivative is the problem of finding the maximum or minimum of a given function. When the function is relatively simple, it is usually fairly easy to find the critical points by algebraically solving $f'(x) = 0$ and then testing each such point with either the First or the Second Derivative Test.

All too often, unfortunately, the function $f(x)$ is quite complicated, so the expression for $f'(x)$ may be far too "messy" to allow us to find the roots by hand. For instance, if $f(x)$ were only a sixth-degree polynomial, it is quite unlikely that we could find the five roots of $f'(x) = 0$, at least not in closed form. As we have seen, though, the use of several of the previous programs, namely, either TABLE or INCDEC and then either NEWTON or BISECT, allows us to find the real roots of $f'(x)$ to any desired degree of accuracy. The fact that we have these critical points, however, still does not solve the problem of optimizing the original function. The program MAXMIN is designed to perform the appropriate analysis of the behavior of a function for each critical point. It is based on an application of the First Derivative Test for extrema (maxima or minima) where the values for the derivative are approximated numerically.

Program MAXMIN

```
10    REM MAX-MIN ANALYSIS FOR EXTREMA
20    INPUT "WHAT IS A CRITICAL POINT? ";X
30    DEF FNY(X) = ...
40    LET H = .0001
50    LET F1 = (FNY(X) - FNY(X - H))/H
60    LET F2 = (FNY(X + H) - FNY(X))/H
70    IF F1 < 0 AND F2 > 0 THEN PRINT "MINIMUM
      AT ";X
80    IF F1 > 0 AND F2 < 0 THEN PRINT "MAXIMUM
      AT ";X
90    IF F1 * F2 > 0 THEN PRINT "NEITHER
      MAXIMUM NOR MINIMUM AT ";X
100   END
```

We note that dividing by h at lines 50 and 60 adds no information to the decision, but is included to make the formulas more familiar.

EXERCISE 21

Modify MAXMIN to provide for the case where either F1 or F2 is zero. In particular, change the value of h to a smaller one and try again.

EXERCISE 22

Modify MAXMIN to handle the case where the critical point x is at either the left or right endpoint of a closed interval $[a, b]$. Compare the value at a to the value at a point slightly larger than a, and similarly at b.

EXERCISE 23   Modify MAXMIN to handle the case where the critical point x is within a distance of h of either endpoint.

EXAMPLE 3.8   To illustrate the use of the sequence of programs involved in solving an optimization problem, suppose we consider the polynomial $f(x) = x^6 - 11x^4 + 15x^3 - 3x - 19$ on the interval $[-10, 10]$. The program TABLE applied to $f'(x)$ with step $h = .5$ yields the following results:

> f' changes sign between -3.5 and -3
>
> f' changes sign between $-.5$ and 0
>
> f' changes sign between 0 and $.5$
>
> f' changes sign between 1 and 1.5
>
> f' changes sign between 1.5 and 2

Since $f'(x)$ is a fifth-degree polynomial, we realize that we have located all five of its roots. In turn, we next approximate each of the five roots by the midpoint of the interval in which it lies and supply each of these values to NEWTON to zero in on the critical points. The results are as follows:

X0 = -3.25	X0 = -.25	X0 = .25	X0 = 1.25	X0 = 1.75
-3.1338754	-.2338689	.3104948	1.2346667	1.8228391
-3.1183062	-.2331693	.3080079	1.2346255	1.8091687
-3.1180434	-.2331619	.3080056		1.8085750
-3.1180433				1.8085740

When each of these five critical values is in turn supplied as input to the program MAXMIN, we finally obtain that there are minima at -3.1180433, $.3080056$, and 1.8085740 and that there are maxima at $-.2331619$ and 1.2346255.

EXERCISE 24   Modify MAXMIN so that it also prints out the value for the maximum or minimum.

It is worth noting that it is certainly possible to write a single program that does all of the above work, so that all the user need do is supply the desired function and the interval. In a practical situation, this would definitely be desirable. However, by removing all the intermediate steps from sight, all that remains is a black box that does the work and simultaneously eliminates all the educational advantages of using the computer. It is therefore recommended that the student be content with the more cumbersome procedure of using the three programs in succession. This is the same line of thought needed to solve the problem with or without the computer.

EXERCISE 25   Write a program similar to MAXMIN to test for extrema using a numerical form of the second derivative test.

EXERCISE 26   Devise a modification of the procedures used above to locate all roots of $f''(x)$ and so determine which, if any, are points of inflection for an arbitrary function $f(x)$.

SECTION 3.6 PROBLEMS

Use the sequence of programs TABLE, then BISECT or NEWTON, and finally MAXMIN (or any modifications of them) to find all local extrema for the following functions. Use this information to sketch the graph of each function.

1 $f(x) = 2x^3 + x^2 - 20x + 1$

2 $f(x) = x^4 - 8x^2 + 1$

3 $f(x) = x^{4/3} + 4x^{1/3}$

4 $f(x) = x^2(\sqrt[3]{x^2 - 4})$

5 $f(x) = x^2/(x^2 - 4)$

6 $f(x) = (x^2 - 1)^{1/3}$

7 $f(x) = x^8 - 11x^6 + 5x^4 + 2x + 4$

8 $f(x) = (x^8 - 11x^6 + 5x^4 + 2x + 4)^{1/2}$

9 $f(x) = x^{3/5} - x^{1/5}$

10 $f(x) = x + \cos x$ on $[0, 10]$

11 $f(x) = \sin \sqrt{x + 1}$ on $[0, 10]$

12 $f(x) = x^{8/3} + x^{7/4} - 2x$

FOUR
THE INTEGRAL

4.1
THE DEFINITE INTEGRAL AND
THE RIEMANN SUM

In the present chapter, we turn our attention to the concept of integration and consider ways in which the computer can be used as a valuable tool to solve problems involving integrals.

According to the Fundamental Theorem of Integral Calculus, we can evaluate a definite integral by

$$\int_a^b f(x)\,dx = F(b) - F(a)$$

where $F(x)$ is any antiderivative of $f(x)$. So long as the given function $f(x)$ is easily integrable, this is an extremely simple procedure. Unfortunately, most of the integrals that arise in practice cannot be handled in closed form, and so we must resort to other methods to evaluate the integral. It is essential we keep in mind, before we proceed, that evaluating a definite integral involves calculating the numerical value for the integral. Thus, in any problem involving complex calculations, we should expect a computer to play an important role.

One of the simplest methods available involves calling upon the definition of the definite integral as the limit of the Riemann Sum:

$$\int_a^b f(x)\,dx = \lim_{\text{all } \Delta x_i \to 0} \sum_{i=0}^{n-1} f(x_i^*)\,\Delta x_i$$

where x_i^* is some point in the ith subinterval $[x_i, x_{i+1}]$ of width Δx_i. For convenience, we assume that the interval $[a, b]$ is uniformly subdivided into n subintervals of equal width $\Delta x = h = (b - a)/n$. See Figure 4.1. Thus, the above formula becomes

$$\int_a^b f(x)\,dx = (b - a) \lim_{n \to \infty} \sum_{i=0}^{n-1} \frac{f(x_i^*)}{n}$$

As we have seen before, the computer cannot evaluate a limit exactly.

FIGURE 4.1

Riemann Sum with $n = 16$ subdivisions

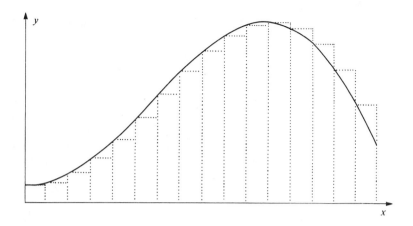

However, if the value for n is taken sufficiently large, then the value of the expression

$$(b - a) \sum_{i=0}^{n-1} \frac{f(x_i^*)}{n}$$

produces an approximation to the value of the definite integral to any desired degree of accuracy.

To simplify the computations as much as possible, we use the left-hand endpoint of each subinterval as the indicated point x_i^* in that sub-interval. That is, we take $x_i^* = x_i$ for each i. Thus, the corresponding approximation to the definite integral is given by

$$\int_a^b f(x)\,dx \approx (b - a) \sum_{i=0}^{n-1} \frac{f(x_i)}{n}$$

for n sufficiently large. This result can be programmed easily for computer evaluation of the value of the definite integral. The result is given in the following program, RIEMANN.

Program RIEMANN

```
10   REM EVALUATION OF DEFINITE INTEGRALS VIA
     RIEMANN SUMS
20   DEF FNY(X) = ...
30   INPUT "WHAT IS THE INTERVAL? ";A,B
40   INPUT "HOW MANY SUBDIVISIONS? ";N
50   LET H = (B - A)/N
60   FOR X = A TO B - H STEP H
70   LET R = R + FNY(X)
80   NEXT X
90   LET R = R * H
100  PRINT "USING ";N; " SUBDIVISIONS, THE
     RIEMANN SUM YIELDS ";R
110  END
```

Before using this program, it is important to realize that the way the computer handles the arithmetic involved will often introduce an error. All numerical calculations are performed in a base 2 system on a computer. Thus, when the computer calculates the value for $h = (b - a)/n$ and then converts it to base 2 form, there are always some round-off errors present. If the round-off happens to form a number larger than the actual value for h, then the program never includes the final term corresponding to $f(b - h)$. It does include the value when $x = b - 2h$; however, the NEXT X will be larger than $b - h$, and so the computer will skip over that term and exit the loop. In order to avoid this potential pitfall, line 60 should be changed to

```
60    FOR X = A TO B - H/2 STEP H
```

Notice that this provides for values of x of $a, a + h, a + 2h, \ldots,$ $a + (n - 1)h = b - h$. However, even if h is rounded upwards, the value computed for $b - h$ definitely falls within the range for the loop. The following value of $x = b$ is definitely beyond the upper limit, thus providing the necessary exit from the loop. The same procedure—forcing the upper limit upwards to ensure that all desired values of x are used—will occur in subsequent programs involving numerical integration.

As an example, suppose we consider the definite integral

$$\int_0^1 \sqrt{x^3 + 1} \, dx$$

which cannot be integrated in closed form by any elementary method. If we now apply the program RIEMANN with a variety of different values for n (see Figure 4.2), then we obtain the following set of values:

$n = 10$	R = 1.13304
$n = 30$	R = 1.11845
$n = 50$	R = 1.11563
$n = 100$	R = 1.11353
$n = 500$	R = 1.11186
$n = 1000$	R = 1.11166
$n = 5000$	R = 1.11149
$n = 10000$	R = 1.11147
$n = 20000$	R = 1.11146

Based on these figures, we see that the value for this definite integral can be taken to be approximately 1.11146. However, note that this involved a considerable amount of computer time.

Moreover, from the above tabulation, clearly the degree of accuracy increases with the number of subdivisions, n. However, when n is taken to be large, we may begin to push the computational limits of the computer in the sense that the time involved in performing the calculation becomes long in the human sense. Thus, with most of the smaller and slower microcomputers, the use of a large value for n will lead to an excruciatingly long wait for a response. As a consequence, it is probably a good idea to use the pro-

FIGURE 4.2

Riemann Sum applied to
$f(x) = \sqrt{x^3 + 1}$ on [0, 1] with
$n = 30$

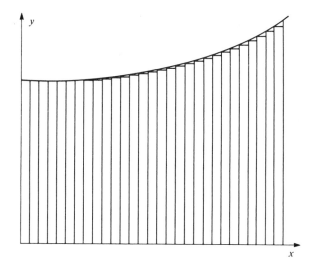

gram with a variety of values for n to assess both the rate of convergence and the time necessary to perform the calculations. Note, though, that a value of n that works for one function may not serve well for another.

| EXERCISE 1 | Modify the program RIEMANN to use the right-hand endpoints for x_i^* instead of the left-hand endpoints. |

| EXERCISE 2 | Modify RIEMANN to use the midpoint of each subinterval for x_i^*. |

| EXERCISE 3 | Use the random number generator RND to select a random number for x_i^* in each subinterval. Use $x_i^* = x_{i+h} \cdot$ random number. |

| EXERCISE 4 | Compare the results obtained with the modifications of RIEMANN in Exercises 1 through 3 when they are applied to $f(x) = \sqrt{x^3 + 1}$ on [0, 1] with $n = 1000$. |

In the following two sections, we will examine several alternative methods for evaluating definite integrals using approximation techniques that are considerably faster and more efficient than the Riemann Sum approach.

SECTION 4.1 PROBLEMS

Use the program RIEMANN or one of your modifications of it to find an approximate value for each of the following definite integrals. In each case, vary the number of subdivisions to see the effects on the accuracy of the approximation. Also, keep track of how long the program RUNs in each case for later comparisons. Wherever possible, compare the numerical results to those obtained by actually evaluating the definite integral.

1 $\displaystyle\int_0^4 (2x + 3)\, dx$

2 $\displaystyle\int_1^4 (3x^2 + 5)\, dx$

3 $\displaystyle\int_1^2 (x^3 + 1)\, dx$

4 $\displaystyle\int_0^2 x\sqrt{8 + 7x^2}\, dx$

5 $\displaystyle\int_0^2 \frac{1}{4 + x^2}\, dx$

6 $\displaystyle\int_0^6 \frac{1}{\sqrt{4 - x^2}}\, dx$

7 $\displaystyle\int_1^3 \frac{x}{x^4 + 1}\, dx$

8 $\displaystyle\int_0^{\pi/2} \sqrt{1 + \sin x}\, dx$

9 $\displaystyle\int_0^{\pi/2} \frac{1}{\sqrt{1 + \sin x}}\, dx$

10 $\displaystyle\int_1^3 x^x\, dx$

11 $\displaystyle\int_0^{\pi/2} (\sin x + \cos x)^{1/3}\, dx$

12 $\displaystyle\int_0^{\pi/2} \sqrt{9 \sin^2 x + 4 \cos^2 x}\, dx$

4.2
THE TRAPEZOID RULE

In the previous section, we employed the Riemann Sum to approximate the value for the definite integral of $f(x)$ on $[a, b]$. Geometrically, this is equivalent to approximating the area of each slice by a rectangle of height $f(x_i^*)$. (See Figure 4.1.) Unfortunately, unless the function is very flat, the resulting approximation is a poor one, and, as a result, we must take an extremely large number of subdivisions n to obtain a reasonably accurate result.

Certainly it would be preferable to find a better approximation — that is, a shape that is a better fit to the curve itself. The *Trapezoid Rule* provides such a better fit where the area of each strip is approximated by the area of a trapezoid, as seen in Figure 4.3 below. The corresponding formula

$$\int_a^b f(x)\, dx \approx \frac{1}{2}\Delta x\{f(x_0) + 2[f(x_1) + f(x_2) + \cdots + f(x_{n-1})] + f(x_n)\}$$

$$= \frac{(b - a)}{2n}\{f(x_0) + 2[f(x_1) + f(x_2) + \cdots + f(x_{n-1})] + f(x_n)\}$$

is derived in detail in all calculus texts, so we will not bother to reproduce the derivation here.

In general, the use of the Trapezoid Rule produces a far more accurate approximation to the value of the definite integral for a given number

FIGURE 4.3

Trapezoid Rule applied to
$f(x) = \sin x$ on $[0, 3]$ with
$n = 9$

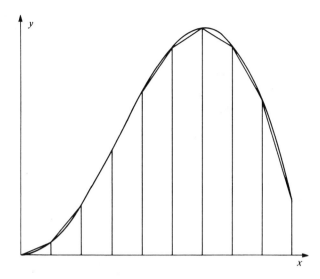

of subdivisions n than the Riemann Sum does. Moreover, the method is an exceptionally appropriate one to use on the computer, as is done in the program TRAP below.

Program TRAP

```
10   REM EVALUATION OF DEFINITE INTEGRALS VIA
     THE TRAPEZOID RULE
20   DEF FNY(X) = ...
30   INPUT "WHAT IS THE INTERVAL? ";A,B
40   INPUT "HOW MANY SUBDIVISIONS? ";N
50   LET H = (B - A)/N
60   LET T = FNY(A) + FNY(B)
70   FOR X = A + H TO B - H/2 STEP H
80   LET T = T + 2 * FNY(X)
90   NEXT X
100  LET T = T * H/2
110  PRINT "USING ";N; " SUBDIVISIONS, THE
     TRAPEZOID RULE YIELDS ";T
120  END
```

Before we illustrate the use of this program, several comments are in order. First, as with the program RIEMANN, we must anticipate the problem of round-off error that can cause an exit from the loop before the last point $x = b - h$ is reached. This is avoided by using the upper value on x of $b - h/2$ instead of $b - h$ at line 70. Second, since all terms except $f(a)$ and $f(b)$ are doubled, it makes sense to handle them separately. Third, it is easier and faster to multiply the final sum for t by the factor $h/2 = \Delta x/2$ at line 100 rather than to multiply each term by it individually within the loop.

EXAMPLE 4.1   We will apply the Trapezoid Rule to the same definite integral

$$\int_0^1 \sqrt{x^3 + 1}\, dx$$

that was treated in the last section to compare the effectiveness of the present method with that of the Riemann Sum approach.

n	Trap	Riemann
10	1.11233239	1.13304
30	1.11154619	1.11845
100	1.11145681	1.11353
200	1.11145018	1.11749
1000	1.11144805	1.11166

Based on these figures and the view shown in Figure 4.4 with $n = 10$, we see that the degree of accuracy possible with the Trapezoid Rule is much greater than that with the Riemann Sum for the same value of n. In fact, in this example, the Trapezoid Rule gave greater accuracy with 30 subdivisions than the Riemann Sum gave with 1000.

EXERCISE 5   Combine the two programs RIEMANN and TRAP into a single program to handle both methods.

Despite the fact that the Trapezoid Rule gives greater accuracy with fewer subdivisions, there is still the question of how large n must be to achieve a desired level of accuracy. In Example 4.1, we essentially experi-

FIGURE 4.4

Trapezoid Rule applied to $f(x) = \sqrt{x^3 + 1}$ on [0, 1] with $n = 10$

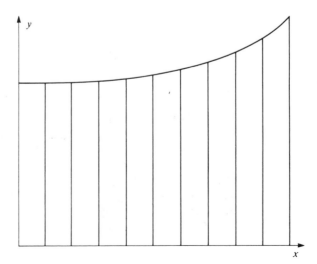

mented with different values of n and just compared the corresponding approximations. When a set of digits matched in succeeding approximations, we naturally concluded that they were correct digits. However, this is at best a haphazard approach and one that should not be relied on if a more systematic procedure is available. Fortunately, such a method does exist for estimating the error in the approximation, and we will illustrate it below as it applies to Example 4.1.

Suppose now that the function $f(x)$ is such that

$$|f''(x)| \leq m_2$$

for all x in the interval $[a, b]$. Thus, m_2 is an upper bound on the absolute value of the second derivative. It turns out that the maximum possible error that can occur when using the Trapezoid Rule with n subdivisions is given by

$$E \leq \frac{m_2(b - a)^3}{12n^2} = E_{\text{max}}$$

The derivation of this result is beyond the scope of this book, and we refer the interested reader to Cheney and Kincaid's book, *Numerical Mathematics and Computing* (Monterey, Calif.: Brooks/Cole Publishing Company, 1985) for the details.

If we examine this error estimate, we see that it is inversely proportional to n^2. Thus, as n increases, the maximum error decreases quickly. Further, we note that in any particular situation the actual error E is likely to be considerably smaller than this estimate. Moreover, the primary difficulty in using such an estimate often lies in finding the upper bound on $|f''(x)|$.

Suppose we now apply this error estimate to Example 4.1 with $f(x) = \sqrt{x^3 + 1}$ on the interval $[0, 1]$. We first find that

$$f'(x) = \frac{3x^2(x^3 + 1)^{-1/2}}{2}$$

so that after some simplification

$$f''(x) = \left(\frac{3}{4}\right)(x^4 + 4x)(x^3 + 1)^{-3/2}$$

We now have to maximize the absolute value of this expression.

We apply the usual calculus techniques to this optimization problem. We differentiate $f''(x)$ and eventually obtain

$$f'''(x) = -\left(\frac{3}{8}\right)(x^6 + 20x^3 - 8)(x^3 + 1)^{-5/2}$$

The only critical points on $[0, 1]$ correspond to the roots of $x^6 + 20x^3 - 8 = 0$ since the denominator $(x^3 + 1)^{-5/2}$ does not become 0 in this interval. Normally we cannot solve for the roots of a sixth-degree polynomial in closed form, but in this case there is a way. We set $u = x^3$, so that we have to solve the quadratic equation

$$u^2 + 20u - 8 = 0$$

whose roots are $u = x^3 = -10 \pm 6\sqrt{3}$. The solution in the interval is approximately .73205081, and it is fairly easy to show that this corresponds to a relative maximum for $f''(x)$ using the first derivative test. Thus, $f''_{max} = 1.467889$. Moreover, the relative minima occur at the endpoints and yield $f''(0) = 0$, $f''(1) = 1.325825$. Therefore, $m_2 = \max |f''(x)| = 1.467889$.

An alternative procedure to approximate this maximum can be implemented using a simple search routine on the computer. We will use M1 to represent the minimum of a function and M2 to represent the maximum.

Program SEARCH

```
10    DEF FNA(X) = ...
20    INPUT "WHAT IS THE INTERVAL? ";A,B
30    INPUT "HOW MANY TEST POINTS? ";N
40    LET M1 = FNA(A): M2 = M1
50    FOR X = A TO B STEP (B - A)/N
60    IF FNA(X) < M1 THEN M1 = FNA(X)
70    IF FNA(X) > M2 THEN M2 = FNA(X)
80    NEXT X
90    PRINT "THE MINIMUM IS ";M1; " AND THE
      MAXIMUM IS ";M2
100   END
```

If we apply this program to $f''(x)$ with $n = 200$ test points, we obtain a value of $m_2 = 1.4679$ that is in close agreement with the value we obtained in the closed form solution above. To account for the possibility that we might miss the actual maximum between two of the test points, however, it is a good idea to use a slightly larger value for the maximum, say $m_2 = 1.5$ in this case.

Suppose now that we use $|f''(x)| \le m_2 = 1.5$. The corresponding maximum error is given by

$$E \le \frac{1.5(b - a)^3}{12n^2} = \frac{.125}{n^2} = E_{max}$$

Therefore, with $n = 10$, the maximum possible error would be .00125; with $n = 30$, it is .00013889. Further, with $n = 100$, the maximum error is .0000125, and we can be certain that the value in the above table is correct to the fourth decimal place. With $n = 1000$, the error is at most .000000135, and we are assured of six significant digits.

Moreover, from the error estimate, we can also find the number of subdivisions n needed to achieve a desired level of accuracy for any given function $f(x)$ on any interval $[a, b]$. As before, let m_2 be the absolute maximum of $f''(x)$ on the interval. Since

$$E_{max} = \frac{m_2(b - a)^3}{12n^2}$$

we easily obtain

$$n^2 = \frac{m_2(b - a)^3}{12E_{max}}$$

so that

$$n = \left[\frac{m_2(b - a)^3}{12E_{max}}\right]^{1/2}$$

Suppose that we want four-place accuracy using the Trapezoid Rule, so that $E_{max} = .00005$. Therefore, using the values $m_2 = 1.5$, $a = 0$, and $b = 1$, we obtain

$$\left[\frac{m_2(b - a)^3}{12E_{max}}\right]^{1/2} = (2500)^{1/2} = 50$$

As a result, we see that any value of $n \geq 50$ will give the desired level of accuracy with the Trapezoid Rule.

SECTION 4.2 PROBLEMS

Use program **TRAP** to find an approximate value for each of the 12 definite integrals in the problem set for Section 4.1. In each case, vary the number of subdivisions to see the effects on the accuracy of the approximation. Estimate the maximum error. Keep track of how long **TRAP** takes in each case, and compare this time to the comparable time when **RIEMANN** was used.

4.3
SIMPSON'S RULE

In the present section, we consider a still more efficient method for evaluating a definite integral: Simpson's Rule. The underlying idea is that in order to obtain a good fit to approximate a curve, another curve is preferable to the straight line used in the Trapezoid Rule. One of the simplest curves available is a parabola, and, as we noted in Section 3.2, a parabola is determined by any three noncollinear points. Therefore, to use a quadratic instead of a linear approximation to a curve, we require a set of three points on the curve. Thus, if we subdivide an interval $[a, b]$ into n subintervals, we must deal with two of the strips at a time to involve three points on the curve, as shown in Figure 4.5. Furthermore, if the subdivisions are taken two at a time without duplication, then the number of strips n must be an even number.

The formula for Simpson's Rule is given by

$$\int_a^b f(x)\,dx \approx \frac{\Delta x}{3}[f(x_0) + 4f(x_1) + 2f(x_2) + 4f(x_3) + 2f(x_4) + \cdots$$
$$+ 4f(x_{n-1}) + f(x_n)]$$

where $\Delta x = (b - a)/n$ and n is even. This formula is derived in almost all

FIGURE 4.5

Simpson's Rule with $n = 4$ subdivisions

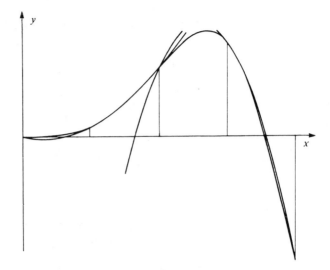

calculus texts and so is not done here. In general, the results using Simpson's Rule are much more accurate than those using the Trapezoid Rule, which in turn are more accurate than those produced by the Riemann Sum. In other words, in order to achieve a given degree of accuracy (a certain number of decimal places), we usually need a much smaller value for n with Simpson's Rule than with either of the other two methods.

The program SIMP is an easy way to implement the use of Simpson's Rule on the computer.

Program SIMP

```
10    REM EVALUATION OF DEFINITE INTEGRALS
      USING SIMPSON'S RULE
20    DEF FNY(X) = ...
30    INPUT "WHAT IS THE INTERVAL? ";A,B
40    INPUT "HOW MANY SUBDIVISIONS? ";N
50    LET H = (B - A)/N
60    LET S = FNY(A) + FNY(B)
70    FOR X = A + H TO B - H/2 STEP 2 * H
80    LET S = S + 4 * FNY(X)
90    NEXT X
100   FOR X = A + 2 * H TO B - 3 * H/2 STEP 2
      * H
110   LET S = S + 2 * FNY(X)
120   NEXT X
130   LET S = H * S/3
140   PRINT "USING ";N; " SUBDIVISIONS,
      SIMPSON'S RULE YIELDS ";S
150   END
```

We note that line 60 is used to take care of the endpoints of the interval where the function's values are counted only once. The loop from line 70 to line 90 accounts for all the odd-numbered terms, where the value of the function is multiplied by 4. The loop from line 100 to line 120 accounts for the even-numbered terms, where the value of the function is doubled. Finally, to increase the speed of the computations, the factor of $\Delta x/3$ is not used for each term individually, but is applied at line 130 only after all previous calculations have been completed.

EXERCISE 6

Since Simpson's Rule requires that the number of subdivisions n be even, the value given to the computer for n should be checked. Modify the program SIMP to determine whether or not n is even by comparing N/2 with INT(N/2).

EXAMPLE 4.2

We once more consider the definite integral

$$\int_0^1 \sqrt{x^3 + 1}\, dx$$

When we applied the Trapezoid Rule to this integral in the last section, we obtained a value of 1.11144805 for the answer using 1000 subdivisions. We now apply SIMP to this integral to produce the values shown in Table 4.1. Incredibly, Simpson's Rule has produced an answer correct to five decimal places with only $n = 10$ subdivisions. The Trapezoid Rule required approximately $n = 100$ to achieve the same level of accuracy.

Admittedly, Simpson's Rule does not always produce such an amazingly good approximation this quickly. It clearly depends on the function in question. If the function has a shape that is closely approximated by a parabola (or a series of parabolas), then we can expect this type of accuracy. If the curve is extremely flat, however, then a parabola is not a particularly good fit, and Simpson's Rule can be considerably less accurate. In fact, for a nearly linear curve, we do better with the Trapezoid Rule. However, for most cases, the use of Simpson's Rule is almost always the best choice among the three.

TABLE 4.1

Simpson's Rule for
$$\int_0^1 \sqrt{x^3 + 1}\, dx$$

n	SIMP
6	1.11143124
8	1.11144270
10	1.11144582
12	1.11144693
20	1.11144784
30	1.11144795
100	1.11144797
1000	1.11144797

<u>EXERCISE 7</u>   Combine the three programs RIEMANN, TRAP, and SIMP into a single program that will allow us to compare the relative effectiveness of the three methods.

We now turn to a consideration of an error estimate for Simpson's Rule that is comparable to the one used in the last section for the Trapezoid Rule. Let m_4 be an absolute upper bound on the fourth derivative of the given function

$$\left|f^{\text{iv}}(x)\right| \le m_4$$

for all x in an interval $[a, b]$. It can be shown that the maximum error possible using Simpson's Rule with n subdivisions is given by

$$E \le \frac{m_4(b - a)^5}{180n^4} = E_{\text{max}}$$

(The interested reader is again referred to Cheney and Kincaid for the derivation.) When we examine this error estimate, we note that it is inversely proportional to the fourth power of n, which accounts for the high degree of accuracy possible with Simpson's Rule.

We can illustrate the use of this result by applying it to the function $f(x) = \sqrt{x^3 + 1}$ from Examples 4.1 and 4.2. In Section 4.2, we saw that

$$f'''(x) = -\left(\frac{3}{8}\right)(x^6 + 20x^3 - 8)(x^3 + 1)^{-5/2}$$

Therefore, after considerable simplification, we obtain

$$f^{\text{iv}}(x) = \left(\frac{9}{16}\right)(x^8 + 56x^5 - 80x^2)(x^3 + 1)^{-7/2}$$

$$= \left(\frac{9}{16}\right)x^2(x^6 + 56x^3 - 80)(x^3 + 1)^{-7/2}$$

While we may find the absolute maximum of this expression by optimization techniques involving $f^{\text{v}}(x)$, the function is getting quite complicated. Therefore, we simply apply the program SEARCH instead and find $m_4 \approx 7.023$. To be on the safe side, since the actual maximum point may have been missed, we use $m_4 = 7.1$. Therefore,

$$E \le .0394444/n^4 = E_{\text{max}}$$

Thus, when $n = 10$, we have $E \le .00000394$ and the corresponding result is accurate to five decimal places. When $n = 20$, $E \le .00000025$ and the approximation is correct to six places.

Note that these results are probably better than we would normally expect to obtain with Simpson's Rule for two reasons. First, the interval $[0, 1]$ is small, so that the term $(b - a)^5$ does not have a large impact. Further, the upper bound on $\left|f^{\text{iv}}(x)\right|$ is relatively small.

As with the error estimate for the Trapezoid Rule, we can also use the above inequality to determine the value of n needed to achieve a desired level of accuracy. We first find

$$n^4 = \frac{m_4(b - a)^5}{180E_{max}}$$

so that

$$n = \left[\frac{m_4(b - a)^5}{180E_{max}}\right]^{1/4}$$

For example, suppose we want to achieve ten-decimal accuracy. We set $E_{max} = .00000000005$ along with $a = 0, b = 1$, and $m_4 = 7.1$ so that the above expression yields

$$(78889000)^{1/4} = 167.59$$

Therefore, any even value of $n \geq 168$ gives the desired accuracy.

SECTION 4.3 PROBLEMS

Use program SIMP or a modification of it to find an approximate value for each of the 12 definite integrals assigned in Section 4.1 and again in 4.2. Estimate the maximum error. Compare the length of time needed by SIMP to achieve a desired degree of accuracy to the time needed by the other two programs.

4.4
MONTE CARLO METHODS

We have seen that many problems in calculus cannot be solved in closed form using the standard methods of calculus. In many instances, we have gotten around such difficulties by introducing appropriate approximations. Thus, in Section 3.2, we used an interpolating polynomial as an approximation to a given function to evaluate its derivative with reasonable accuracy. In Sections 4.1 through 4.3, we approximated the area under a curve (and therefore a definite integral) by considering rectangular strips using horizontal line segments (Riemann Sum), trapezoids using diagonal line segments (Trapezoid Rule), and portions of parabolas (Simpson's Rule), respectively, to evaluate a definite integral.

A variety of altogether different approaches have been developed to solve calculus-type problems that do not involve applying calculus methods to approximations. One particularly fascinating approach is called the Monte Carlo Method and can be used to approximately evaluate definite integrals. As the name suggests, it is based on probabilistic methods such as occur in gambling situations. It can often be implemented easily on a computer using the random number function RND.

Suppose we want to find the area under the curve $f(x) = x^2$ from $x = 0$ to $x = 1$. By integrating, we know that the answer is

$$\int_0^1 x^2\, dx = \frac{x^3}{3}\Big|_0^1 = \frac{1}{3}$$

Consider the graph of this function as shown in Figure 4.6. We see that the region in question is precisely contained in the rectangle $R:[0, 1] \times [0, 1]$, whose area is exactly 1. Suppose we can pick a truly random point anywhere in the rectangle. Then, the chance of it occurring inside the desired region A is 1 out of 3, since the area of A is ⅓ of the total area of R.

Now let us perform this process repeatedly. Each random point has a probability of ⅓ of landing in the region A. Thus, if we repeat this process 30,000 times, say, then we can expect approximately 10,000 of the points to fall inside of A.

It is important to realize that this approach requires large numbers of repetitions. If the process were repeated only 100 times, certainly a disproportionate number of the points might fall inside of A, maybe 50 out of the 100. In the long run, however, as the number of repetitions increases, the chance discrepancies tend to balance out, and the theoretical values are approached asymptotically.

We can implement the Monte Carlo procedure easily in BASIC. Recall that, for most computer systems, RND(0) produces a random number between 0 and 1. In order to determine a point inside the rectangle R, we need two coordinates between 0 and 1. Therefore, we set X = RND(0) and Y = RND(0) and use C to count how many of the random points fall inside the region A; for example, whether $y < x^2$. We do this with the following simple program:

```
10    INPUT "HOW MANY REPETITIONS N? ";N
20    FOR I = 1 TO N
```

FIGURE 4.6

Graph of $f(x) = x^2$ inside box
$[0, 1] \times [0, 1]$

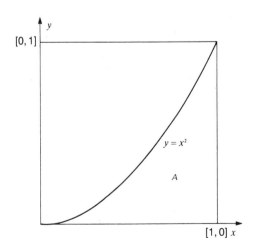

```
30   LET X = RND(0)
40   LET Y = RND(0)
50   IF Y <= X * X THEN C = C + 1
60   NEXT I
70   PRINT "THERE WERE ";C; " POINTS IN THE
     REGION"
80   END
```

In one sample RUN of this program, we obtain a value of $c = 9866$ out of 30,000.

On the other hand, suppose that we did not know that the area of region A was $1/3$. We could conclude from the above process that since 9866 of the 30,000 random points (or 32.8867% of them) fell inside A, then the area of A should be approximately the same proportion of the total area of the enclosing rectangle R. Thus, since the area of R is 1, we would decide that the area of A is approximately 32.8867% of 1 = .328867. Clearly, our conclusion is quite close to the actual value of $1/3$.

This Monte Carlo Method can be extended to find the area under a wide variety of curves $y = f(x)$. For simplicity, suppose that $f(x) \geq 0$ on an interval $[a, b]$. Further, suppose that we can find some number m such that $f(x) \leq m$ on this interval. The number m is known as an upper bound on the function $f(x)$. Note that there are infinitely many possible upper bounds m, since m need not be the actual maximum value for the function. Any value larger than the maximum also works. Based on these assumptions, it follows that the region in question lies entirely within a rectangle $R:[a, b] \times [0, m]$, as shown in Figure 4.7. Further, the area of the rectangle R is simply $(b - a)m$.

We now generate a series of n random points inside this rectangle as follows. The x-coordinate must be between a and b while the y-coordinate must be between 0 and m. Since RND(0) produces a number between 0 and 1, M * RND(0) will be between 0 and m. Further, (B − A) * RND(0) will

FIGURE 4.7

Graph of a function inside box $[a, b] \times [0, m]$

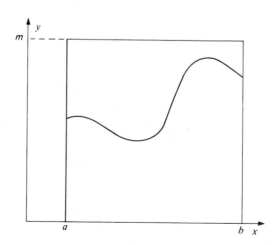

lie between 0 and $b - a$. Therefore, A + (B − A) * RND(0) will lie be-tween a and $a + (b - a) = b$. Moreover, it is also necessary to keep track of how many of the points actually fall under the curve. This can be tested easily, since a point (x_0, y_0) lies under the curve $y = f(x)$ if $y_0 \leq f(x_0)$. Again, we use c to count the actual number of points that satisfy $y_0 \leq f(x_0)$. The proportion c/n times the area of the rectangle, $(b - a)m$, is then taken to be the approximation to the area of the region. This is implemented using the following program.

Program MONTE

```
10    DEF FNA(X) = ...
20    INPUT "WHAT IS THE INTERVAL? ";A,B
30    INPUT "WHAT IS AN UPPER BOUND ON F? ";M
40    INPUT "HOW MANY REPETITIONS? ";N
50    FOR I = 1 TO N
60    LET X = A + (B - A) * RND(0)
70    LET Y = M * RND(0)
80    IF Y <= FNA(X) THEN C = C + 1
90    NEXT I
100   PRINT "THE AREA IS APPROXIMATELY ";C/N
      * (B - A) * M
110   END
```

EXAMPLE 4.3

Apply the Monte Carlo Method to approximate the value of

$$\int_0^\pi \sqrt{1 + \cos^2 x}\, dx$$

An upper bound for $f(x) = \sqrt{1 + \cos^2 x}$ is clearly $m = \sqrt{2}$. On a sample run of MONTE with $n = 10{,}000$, we find that 8599 of the random points fall into the desired region, so that the area under the curve is approximately .8599 times the area of the encompassing rectangle, or

$$.8599m(b - a) = .8599(\sqrt{2})(\pi)$$

$$= 3.8204$$

We note that the major disadvantage in using the Monte Carlo Method is that it takes considerably longer to RUN than any of the previous numeri-cal integration approaches, because the function $f(x)$ has to be evaluated so many times. For example, if we use Simpson's Rule with $n = 1000$ sub-divisions, we would obtain an extremely accurate result and this involves evaluating $f(x)$ at 1001 points. However, to use the Monte Carlo Method, we might need 10,000 or even 100,000 repetitions and this takes proportionately longer. Nevertheless, there are some situations where such probabilistic methods are preferable. The following example illustrates one such case where the Monte Carlo Method provides a different point of view on a well-known fact.

EXAMPLE 4.4

Use the Monte Carlo Method to find an approximate value for π.

Consider the unit circle $x^2 + y^2 = 1$ whose area is precisely π square units. This circle can be inscribed in a square with side 2, as shown in Figure 4.8. The area of the square is 4 square units. The ratio of the two areas is just $\pi/4$.

We now use a slight variation on MONTE that generates random points between -1 and 1 and tests whether $x^2 + y^2 \leq 1$. Using $n = 10,000$, we find that on one sample RUN, the ratio c/n is .78375. Since this represents an approximation to the ratio $\pi/4$, we conclude that $\pi \approx 4(.78375) = 3.1350$. Needless to say, this is not an outstanding approximation, but we can improve it by taking a much larger value for n.

EXERCISE 8

Modify MONTE so that it contains a routine to determine an upper bound on the function f on the given interval. In particular, for a relatively small step size h, test the values of $f(a + ih)$ across the interval and keep track of the largest value m so obtained. To be on the safe side, it might be a good idea to multiply the value you obtain by a small scaling factor, say 1.1, to account for the possibility of missing the actual maximum of the function.

This exercise suggests one way to determine the value of an upper bound m using a computer search. A more certain procedure, of course, is to apply the methods of calculus to the function $f(x)$ to determine its absolute maximum precisely. Of course, when a function becomes sufficiently complicated, the closed form methods may not be particularly easy to apply.

There is still another way to determine, at least approximately, the maximum of a function $f(x)$ on an interval $[a, b]$. Instead of using the systematic search procedure suggested in Exercise 8, where the function's values are checked at the sequence of points $a, a + h, a + 2h, \ldots, b$, we can apply a variation on the Monte Carlo Method. In particular, rather than consider 100 points, say, uniformly distributed across the interval from a to

FIGURE 4.8

Unit circle inscribed in a square
with side 2

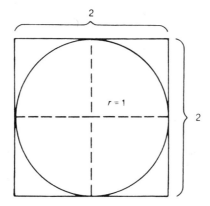

b, we instead can randomly select 100 points between a and b, and check the value of the function at each point. The largest such value then can be used as an approximation to the maximum m. This approach works surprisingly well to give a relatively accurate value for the maximum.

SECTION 4.4 PROBLEMS

Apply the program MONTE or one of your modifications of it to Problems 1 through 12 of Section 4.1.

4.5
THE MEAN VALUE THEOREM FOR INTEGRALS

According to the Mean Value Theorem for Integrals, if $f(x)$ is continuous on a closed interval $[a, b]$, then there exists a point c between a and b such that

$$f(c) = \frac{1}{b - a} \int_a^b f(x)\, dx$$

Geometrically, this means that there is a point c between a and b where the rectangle formed with height $f(c)$ has precisely the same area as under the curve. See Figure 4.9.

As with the Mean Value Theorem for Derivatives, the number c here is guaranteed to exist, but it is rarely possible to solve for it in closed form. However, we saw in previous sections that where a certain quantity is known to exist (the number c in the Mean Value Theorem for Derivatives, Section 3.5, and the value of delta corresponding to a given epsilon, Section 2.7), we can determine that quantity approximately using the computer.

FIGURE 4.9

Graphical representation of the Mean Value Theorem for Integrals

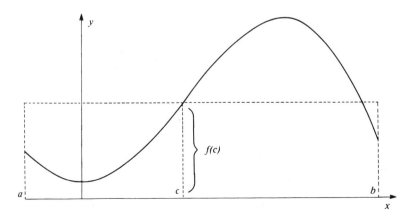

The same is true of the number c whose existence is guaranteed by the Mean Value Theorem for Integrals.

The procedure involved in the corresponding program has two distinct stages. First, the definite integral, I, must be evaluated to produce the value for $f(c)$ guaranteed by the theorem. This can best be done using Simpson's Rule. Once the value is found, we consider the function $g(x) = f(x) - I/(b - a)$. The number c corresponds to a root of this function where $g(c) = f(c) - I/(b - a) = 0$, and such a root can be found using Newton's Method. The initial estimate for this root can best be selected as the midpoint of the interval $[a, b]$.

The resulting program is fairly lengthy and so is not given here in its entirety. Rather, a canned version is available under the name MVTINT. To use it, we must define not only the given function FNY(X) but also the first derivative FND(X), since it is required in the use of Newton's Method.

EXAMPLE 4.5

Consider the function $f(x) = \sin x$ on the interval $[0, \pi]$. When the program MVTINT is applied to this function on this interval, it yields a value of $c = .69010779$.

It is possible to calculate the value for c directly in this case, and we will do so for comparison. According to the Mean Value Theorem for Integrals,

$$f(c) = \sin c = \frac{1}{\pi} \int_0^\pi \sin x \, dx = \frac{-1}{\pi} \cos x \Big|_0^\pi$$

$$= \frac{2}{\pi} = .636620$$

Therefore, using the inverse sine key on a calculator, we obtain $c = \sin^{-1}(.636620) = .69010739 \ (= 39.54024°)$. We observe that the program provides accurate results, at least for this case.

EXAMPLE 4.6

Consider the function $f(x) = \sqrt{x} + \sin x$ on the interval $[0, \pi]$. This function cannot be integrated in closed form, so we have no alternative to the use of the computer to determine c. The program MVTINT now yields $c = 1.1386261$. With this value for c, we note that $f(c) = 1.43062399$ agrees with $I/(b - a)$ to 8 decimal places.

EXERCISE 9

Write your own version of the program MVTINT that calculates c for any function according to the Mean Value Theorem for Integrals. Use either the Bisection Method or Newton's Method to determine the root of $g(x)$, after having calculated the value of the definite integral with Simpson's Rule with $n = 1000$ subdivisions.

SECTION 4.5 PROBLEMS

Use either the prepared program MVTINT (if available) or a succession of previous programs such as SIMP followed by either BISECT or NEWTON to determine a number c satisfying the Mean Value Theorem for Integrals for each of the following definite integrals. Note that you will have already evaluated many of these integrals in the previous problem sets.

1 $\displaystyle\int_{-1}^{2} (x^2 + 5x - 4)\, dx$

8 $\displaystyle\int_{1}^{3} \frac{x}{x^4 + 1}\, dx$

2 $\displaystyle\int_{-1}^{1} (2x^3 - 3x^2 + 4x - 1)\, dx$

9 $\displaystyle\int_{0}^{\pi/2} \sqrt{1 + \sin x}\, dx$

3 $\displaystyle\int_{0}^{3} \sqrt{5x + 1}\, dx$

10 $\displaystyle\int_{0}^{\pi/2} \frac{1}{\sqrt{1 + \sin x}}\, dx$

4 $\displaystyle\int_{0}^{2} x\sqrt{8 + 7x^2}\, dx$

11 $\displaystyle\int_{1}^{3} x^x\, dx$

5 $\displaystyle\int_{0}^{2} \frac{1}{4 + x^2}\, dx$

12 $\displaystyle\int_{0}^{\pi/2} (\sin x + \cos x)^{1/3}\, dx$

6 $\displaystyle\int_{0}^{6} \frac{1}{\sqrt{4 - x^2}}\, dx$

13 $\displaystyle\int_{0}^{\pi/2} \sqrt{9 \sin^2 x + 4 \cos^2 x}\, dx$

7 $\displaystyle\int_{1/2}^{1} \left(1 + \frac{1}{x}\right)^{1/3} \left(\frac{1}{x^2}\right) dx$

4.6
APPLICATIONS OF THE DEFINITE INTEGRAL

There are certain classes of problems, all involving definite integrals, that are common to most calculus courses. Among these are finding the area under a curve, the area between two curves, the volume of a solid of revolution, the length of arc, work, and pressure. In each instance, the problem reduces to evaluating the appropriate definite integral. However, that is often easier said than done: unless great care is taken in choosing the function for a textbook problem, it will probably be impossible to find the volume of a solid — integrating $f(x)^2$ — or the arc length — integrating $\sqrt{1 + f'(x)^2}$ — in closed form. As a consequence, the examples and problems seen in most textbooks are limited and artificial to guarantee that the integration works out.

On the other hand, we now have at our disposal a variety of numerical techniques for evaluating a definite integral. The most effective of these is Simpson's Rule using the program SIMP. In using it, though, we must be careful to define the correct function to the computer for the problem at hand. For example, if the region under the curve $y = f(x)$ is rotated about the x-axis, then the function to be integrated is $\pi f(x)^2$, so that line 20 in the program SIMP must be

```
20   DEF FNY(X) = 3.14159 * ( F(X) )^2
```

If the region is rotated about the y-axis instead, then the function to be integrated is $2\pi x f(x)$, and so line 20 becomes

```
20   DEF FNY(X) = 2 * 3.14159 * X
     * ( F(X) )
```

Finally, if the arc length is to be found, then the integrand is $\sqrt{1 + f'(x)^2}$, so that line 20 then becomes

```
20   DEF FNY(X) = SQR(1 + ( F'(X) )^2)
```

EXAMPLE 4.7

Find the length of arc of the curve $y = f(x) = x^4$ from $x = 0$ to $x = 5$.
Since $f'(x) = 4x^3$, the arc length is given by

$$l = \int_0^5 \sqrt{1 + 16x^6}\, dx$$

which cannot be evaluated in closed form. Therefore, we apply the program SIMP with the function

```
DEF FNY(X) = SQR(1 + 16 * X^6)
```

and use $n = 200$ subdivisions to obtain an answer of 625.66 units.

Often, problems on applications of the definite integral involve curves where x is a function of y, $x = g(y)$. The corresponding formulas are obtained by simply interchanging x and y in the standard formulas for the integrals. At first thought, then, we might believe that we would have to change the entire program SIMP in a similar fashion to have y as the independent variable. However, this is not necessary. The variable x used in the program is just a convenient name tag used by the computer for the independent variable. Therefore, all that we have to do to use the program when x is a function of y is to replace y with x in the definition for the function, as shown in the following illustration.

EXAMPLE 4.8

Find the volume generated when the region between the curve

$$x = g(y) = \sqrt{y} + \frac{y}{y + 2}$$

and the y-axis for $y = 1$ to $y = 4$ is rotated about the y-axis.

Since the region is given in terms of y and is rotated about the y-axis, the solid so generated would be partitioned by disks. Therefore, the volume of the solid is given by

$$V = \pi \int_1^4 \left(\sqrt{y} + \frac{y}{y + 2} \right)^2 dy$$

To evaluate this integral using the program SIMP (it cannot be done easily in closed form), we simply define the function

```
DEF FNY(X) = 3.14159 * (SQR(X)
+ X/(X + 2))^2
```

by replacing the independent variable y with the "dummy" variable x for the computer. If we take $n = 1000$ subdivisions, the value for the volume of this solid then is determined to be 42.63837 cubic units.

EXAMPLE 4.9

Find the length of arc of one arch of the sine curve.

We need to find the arc length of $f(x) = \sin x$ from $x = 0$ to $x = \pi$. Therefore, $f'(x) = \cos x$, so that

$$l = \int_0^\pi \sqrt{1 + \cos^2 x} \, dx$$

However, this is precisely the definite integral treated in Example 4.3 in Section 4.4, so that MONTE has already given us an approximate answer of 3.8204. The exact answer, correct to six decimal places, is 3.820194.

SECTION 4.6 PROBLEMS

Apply the program SIMP or a modification of it to obtain an approximate solution to each of the following problems. When necessary, use the program TABLE to determine the geometric orientation of curves and then either program BISECT or NEWTON to locate points where the curves cross the appropriate axis or cross one another.

1 Find the area between the following pairs of curves between the indicated points.

(a) $f(x) = \dfrac{1}{\sqrt{1 + x^3}}$ and $g(x) = \dfrac{1}{\sqrt{1 + x^4}}$ from $x = 2$ to $x = 3$

(b) $f(x) = \sin(x^2)$ and $g(x) = \dfrac{x}{x^3 + 1}$ between $x = 0$ and $x = 1$

(c) $f(y) = 2^{(y^2)}$ and $g(y) = \sqrt{1 + \cos y}$ from 0 to π

2 Find the volume of the solid of revolution obtained when the given curve is rotated about the indicated axis.

(a) $f(x) = \dfrac{\sin x}{x}$ about the x-axis for x between 1 and $\dfrac{\pi}{2}$

(b) $f(x) = \tan x$ about the y-axis from 0 to $\dfrac{\pi}{4}$

(c) $f(y) = (1 + y + y^2)^{3/5}$ about the y-axis from 0 to 1

3 Find the volume of the solid of revolution obtained when the region between the given curves is rotated about the indicated axes.

(a) $f(x) = \dfrac{\cos x}{x}$ and $g(x) = \dfrac{x}{\cos x}$ about the x-axis from $x = \dfrac{\pi}{6}$ to

$x = \dfrac{\pi}{3}$

(b) the same region rotated about the y-axis

4 Find the arc lengths of the following curves.

(a) $f(x) = x^2$ from $x = 0$ to $x = 1$

(b) $f(x) = x^3$ from $x = 1$ to $x = 4$

(c) $y = \sin x$ from $x = 0$ to $x = \pi$

(d) $y = x^{5/3}$ from $x = 1$ to $x = 4$

FIVE
THE
TRANSCENDENTAL FUNCTIONS

5.1
EXPONENTIAL AND
LOGARITHMIC FUNCTIONS

Thus far, we have seen how many of the topics in calculus can be treated using a computer. In many cases, the computer has been a useful tool, allowing us to solve problems that ordinarily cannot be done by hand or in closed form. In other instances, computer programs have provided additional insights into and understanding of various calculus concepts. In the process, we essentially have treated the topics in the usual order in which they are introduced in most calculus courses. As a result, most applications so far have involved polynomials, rational functions, and root functions.

At this point, we turn our attention to the study of exponential and logarithmic functions, with particular emphasis on the use of e as the base for the natural logarithm, $\ln x$, and the associated exponential function e^x. Since these two functions are so vitally important throughout mathematics, the BASIC language has built-in functions to evaluate them, as discussed previously in Chapter 1. In order to obtain the value of e^x for any value x we simply use EXP(X). Similarly, to obtain the value of $\ln x$ for any x we just type LOG(X).

The base e is the only base that the computer recognizes implicitly. If there is need for any other base, say $b = 10$, then it is necessary to construct the corresponding functions. For exponential functions, it is simple: we merely use B^X. For logarithms, however, things are somewhat more complicated if we want to use a base b other than e. If we write $y = \log_b x$, then we know that $x = b^y$, so that

$$\ln x = \ln b^y = y \ln b = \log_b x \ln b$$

Therefore, it follows that

$$\log_b x = \frac{\ln x}{\ln b}$$

and every time we refer to such a logarithm on the computer, we have to include the full expression

```
LOG(X)/LOG(B)
```

Having introduced these two new classes of functions — exponential and logarithmic — we should realize that all the previous concepts and methods of calculus apply to them, as well as to the types of functions previously considered. Thus, for example, we should be able to find the maxima and minima of functions involving exponentials. We should be able to evaluate definite integrals involving logarithms. We should be able to find the limits of such functions, and so forth. Furthermore, all the computer programs we have used to deal with such topics will also apply to these classes of functions.

EXAMPLE 5.1

Generate a table of values for the function $f(x) = \ln x$.

We use the program TABLE from Section 2.1 with the following change:

```
20   DEF FNY(X) = LOG(X)
```

and the desired table is easily generated. Appropriate headings can be added very simply.

EXERCISE 1

Modify TABLE so that the output is rounded off to four decimal place accuracy by using the INT function in the form INT(10000 * X) + .5/ 10000.

EXAMPLE 5.2

Evaluate

$$\lim_{x \to 1} \frac{\ln x}{x - 1}$$

We use the program LIMFN from Section 2.4 with

```
FNY(X) = LOG(X)/(X - 1)
```

If we use the sequence $h = 1/2^n$ that converges to zero moderately fast, then we obtain the results shown in Table 5.1. Based on these results, we conclude that the limit is probably 1.

EXAMPLE 5.3

Apply the delta-epsilon definition of limit to the function in Example 5.2. In particular, find a delta corresponding to epsilon equal to .073, .0073, and .00073.

We use the program DELEPS with $a = 1$ and $L = 1$. When ε is .073, the program responds with $\delta = .1$. When ε is .0073, the program produces $\delta = .0125$. Finally, when ε is .00073, the program calculates $\delta = .00078125$.

TABLE 5.1

Values for $\lim_{x \to 1} \ln x/(x - 1)$

A - H	FNY(A - H)	A + H	FNY(A + H)
.5	1.38629	1.5	.810931
.75	1.15073	1.25	.892575
.875	1.06825	1.125	.942266
.9375	1.03261	1.0625	.969995
.96875	1.01595	1.03125	.984693
.984375	1.00788	1.01563	.992272
.992188	1.00389	1.00781	.996119
.996094	1.00193	1.00391	.998049
.998047	1.00087	1.00195	.99898
.999023	1.00035	1.00098	.999572
.999512	.999911	1.00049	.999614
.999756	.999191	1.00024	1.00012

EXAMPLE 5.4

Find all maxima and minima for the function

$$f(x) = (x^3 - 4x^2 + 1)[\ln(x + 4)]e^{-x}$$

on the open interval $(-1, 9)$.

We first apply the program INCDEC from Section 2.2, which tells us

The function increases from -1 to 0,

The function decreases from 0 to 1.5,

The function increases from 1.5 to 5.5,

The function decreases from 5.5 to 9.

We therefore use the values 0, 1.5, and 5.5 as input to the program NEWTON from Section 3.4 to represent the initial estimates for the roots of the derivative $f'(x)$. We also supply the derivative of the function $f(x)$ here as the given function FNY(X) in the program. The resulting output from NEWTON provides us with critical points at $x = -.094455$, 1.678085, and 5.620985. We then use these values as input to the program MAXMIN in Section 3.6 to find that there is a minimum at $x = -.094455$, a minimum at $x = 1.678085$, and a maximum at $x = 5.620985$. See Figure 5.1 for the graph.

EXAMPLE 5.5

Evaluate the definite integral

$$\int_0^2 \sqrt{1 + e^{-3x}}\, dx$$

We first apply the program RIEMANN from Section 4.1 with $n = 500$ subdivisions to obtain a value of 2.151072 for the integral. We next apply the program TRAP from Section 4.2, with the same number of

FIGURE 5.1

Graph of $f(x) = (x^3 - 4x^2 + 1)$
$\cdot \ln(x + 4) \cdot e^{-x}$ on $(-1, 9)$

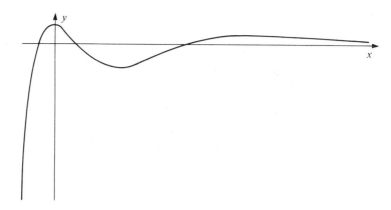

Note: x ranges from -1 to 9 in steps of 1
y ranges from -11.945 to 1.437 in steps of 1.3382

subdivisions, to obtain 2.150246. Finally, we apply the program SIMP from Section 4.3 to obtain 2.150245, which is probably the most accurate.

SECTION 5.1 PROBLEMS

Apply any of the programs from the preceding chapters that are necessary to solve the following problems.

1 Find $\lim\limits_{x \to 0} (e^x - 1)/x$.

2 Find $\lim\limits_{h \to 0} (e^{x+h} - e^x)/h$ for $x = 0$, $x = 1$, $x = -1$.

3 Find $\lim\limits_{x \to 0} \ln(x)/x$.

4 Find $\lim\limits_{x \to \infty} \ln x/e^x$.

5 Find $\lim\limits_{x \to 0^+} x^{1/\ln x}$.

6 Find $\lim\limits_{t \to \infty} 2600(1 - .51e^{-.075t})^3$.

7 Find the derivative of $f(x) = \ln x$ at $x = 1$, $x = 2$, $x = 3$.

8 Find a number c predicted by the Mean Value Theorem for the function $f(x) = xe^{-x}$ on the interval $[0, 2]$.

9 Find the point where $f(x) = x$ crosses $g(x) = e^{-x}$.

10 Find the root of $(1/x) - \ln x = 0$.

11 Find the area under the curve $f(x) = \ln(x + e^x)$ from $x = 0$ to $x = 3$.

12 Find the volume obtained when the region in Problem 11 is rotated about the x-axis; about the y-axis.

5.2
EVALUATING EXPONENTIAL AND LOGARITHMIC FUNCTIONS

In the previous section, we saw how the computer can be used to deal with functions involving exponential and logarithmic functions. It is instructive to see how the values for these functions are actually evaluated by either the computer or a hand-held calculator.

In Section 2.5, we considered an example involving the limit

$$\lim_{x \to \infty} \left(1 + \frac{1}{x} \right)^x$$

that turned out to be approximately 2.718. By this stage in our study of calculus, however, this number should be immediately recognizable as being the first few digits of the transcendental number

$$e = 2.71828\ 1828\ 459045 \dots$$

that is, the base of the natural logarithm system. Therefore, it seems that

$$e = \lim_{t \to \infty} \left(1 + \frac{1}{t} \right)^t$$

where we have introduced the variable t in place of x. Therefore, when t is large, it follows that

$$\left(1 + \frac{1}{t} \right)^t \approx e$$

Alternatively, if we now let $h = 1/t$, so that $t = 1/h$, and if we notice that as t approaches infinity, h must approach zero, then we find that

$$e = \lim_{h \to 0} (1 + h)^{1/h}$$

We note that for an expression such as this, we cannot simply "plug in" $h = 0$; the result would be 1^∞, and this is not a well-defined arithmetic quantity. Consequently, the best we can do is to evaluate the expression for a sequence of values of h that converge to 0 using a program similar to the program LIMFN in Section 2.4.

Now, using the above limit that leads to e, when h is small, we have

$$e \approx (1 + h)^{1/h}$$

so for any real value of x, it follows that

$$e^x \approx (1 + h)^{x/h}$$

If we now apply a modification of LIMFN to this expression, we obtain a sequence of values, depending on the choice of x, that converges to the value of e^x as $h \to 0$. The change needed is based on the fact that the variable h is the one that is changing, while the value used for x is essentially a "dummy" variable—it is not affected by the limit process or by the program.

Program EXLIM

```
10   REM EXPONENTIAL FUNCTIONS VIA THE LIMIT
20   INPUT "WHAT IS THE VALUE FOR X? ";X
30   INPUT "HOW MANY TERMS DO YOU WANT? ";N
40   FOR I = 1 TO N
50   LET H = ...
60   LET P = (1 + H)^(X/H)
70   PRINT I,H,P
80   NEXT I
90   END
```

EXAMPLE 5.6

We will apply the program EXLIM using $x = 1$, since we already know what the result should be, namely $e^1 = e = 2.71828\ldots$ For the sequence h, we choose at line 50

$$H = 1/I^4$$

to illustrate the convergence. The corresponding output is shown in Table 5.2. Based on Table 5.2, we see that the sequence of values is indeed converging to the expected value for e.

While the approach suggested above to calculate values for e^x using the program EXLIM is valid in theory, it is not the approach actually implemented in practice. However, by some relatively simple manipulations, we can obtain a considerably more effective technique. We previously noted that when h is small,

$$e^x \approx (1 + h)^{x/h}$$

TABLE 5.2

Evaluating e using limits

i	h	p
1	1.	2.
2	.0625	2.637929
3	.0123458	2.701690
4	.0039062	2.712991
5	.0016	2.716111
6	7.71605 E-4	2.717235
7	4.16493 E-4	2.717718
8	2.441406 E-4	2.717947
9	1.52416 E-4	2.718065
10	1.E-4	2.718147
11	6.83013 E-5	2.718188
12	4.82253 E-5	2.718265

FIGURE 5.2

Graph of e^x versus graphs of $(1 + h)^{x/h}$ for $h = .5, .25, .125, .0625$

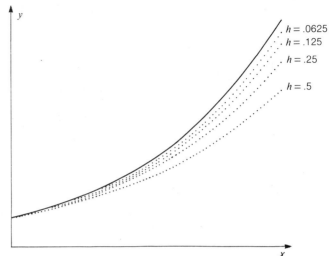

Note: x ranges from 0 to 2 in steps of .2
y ranges from 0 to 7.38906 in steps of .738906

(In Figure 5.2, we show the graphs of $(1 + h)^{x/h}$ as functions of x for different values of h.) The expression on the right can be expanded using the Binomial Theorem to yield

$$
\begin{aligned}
e^x \approx (1 + h)^{x/h} = \ & 1 + \left(\frac{x}{h}\right)h + \left(\frac{x}{h}\right)\left(\frac{x}{h} - 1\right)\frac{h^2}{2} \\
& + \left(\frac{x}{h}\right)\left(\frac{x}{h} - 1\right)\left(\frac{x}{h} - 2\right)\frac{h^3}{6} \\
& + \left(\frac{x}{h}\right)\left(\frac{x}{h} - 1\right)\left(\frac{x}{h} - 2\right)\left(\frac{x}{h} - 3\right)\frac{h^4}{24} + \cdots \\
= \ & 1 + \left(\frac{x}{h}\right)h + \left(\frac{x}{h}\right)\left(\frac{x - h}{h}\right)\frac{h^2}{2!} \\
& + \left(\frac{x}{h}\right)\left(\frac{x - h}{h}\right)\left(\frac{x - 2h}{h}\right)\frac{h^3}{3!} \\
& + \left(\frac{x}{h}\right)\left(\frac{x - h}{h}\right)\left(\frac{x - 2h}{h}\right)\left(\frac{x - 3h}{h}\right)\frac{h^4}{4!} + \cdots \\
= \ & 1 + x + \frac{x(x - h)}{2!} + \frac{x(x - h)(x - 2h)}{3!} \\
& + \frac{x(x - h)(x - 2h)(x - 3h)}{4!} + \cdots
\end{aligned}
$$

where we have introduced the factorial notation $3! = 1 \cdot 2 \cdot 3$, $4! = 1 \cdot 2 \cdot 3 \cdot 4$, and $n! = 1 \cdot 2 \cdot 3 \cdots (n - 1)n$. If we now let $h \to 0$, then the above approximation for e^x approaches

$$e^x \approx 1 + x + \frac{x^2}{2!} + \frac{x^3}{3!} + \frac{x^4}{4!} + \cdots$$

This expression will become familiar later on in the study of infinite series. For now, though, we comment that this is the method that is used in practice to calculate the values for e^x for any value of x. That is, the polynomial expression on the right is simply evaluated for any given value of x using an appropriate number of the terms. We note that the number of terms needed will depend on the size of x: The larger x is, the more terms will be required.

The following program, EXPX, is designed to implement this method of calculating an approximate value for e^x for any given x. We note that the result will always be an approximation, since the precise value for e^x would involve all of the terms indicated in the above expression by the dots, and it is impossible to do this.

Program EXPX

```
10   REM CALCULATION OF EXP(X)
20   INPUT "WHAT IS THE VALUE OF X? ";X
30   INPUT "HOW MANY TERMS? ";M
40   LET A = 1
50   FOR N = 0 TO M
60   LET S = S + A
70   LET A = A * X/(N + 1)
80   PRINT N,S
90   NEXT N
100  END
```

We note that line 70 is used to simplify the calculation of the terms needed in the form of $x^n/n!$. Thus, instead of calculating each of these terms as it is needed, we instead use each term, say $x^3/3!$ for a at one stage, and then multiply it by $x/4$ to form $x^4/4!$ at the next stage of the loop.

EXAMPLE 5.7

We again use $x = 1$ to calculate the value of e, this time with the program EXPX. The resulting output using $m = 9$ is

```
0   1
1   2.
2   2.5
3   2.666667
4   2.708333
5   2.716667
6   2.718055
7   2.718254
8   2.718279
9   2.718282
```

Based on these results, we notice that the process used here with EXPX converges more rapidly than the previous method did using the program EXLIM.

EXERCISE 2

Modify EXPX so that it provides a check for divergence due to round-off errors.

Just as e^x can be expressed as a limit, it is also possible to show that the natural logarithm ln x arises from a limit as well, namely,

$$\ln x = \lim_{h \to 0} \frac{x^h - 1}{h}$$

for any positive real number x. As a result, we can use a modification of the program EXLIM, called LOGLIM, to evaluate the value of ln x for any x using a sequence.

EXERCISE 3

Write the program LOGLIM.

EXAMPLE 5.8

We run the program LOGLIM using $x = 2$, noting the fact that ln 2 = .693147. The corresponding output is found in Table 5.3. If we go beyond this stage, the sequence begins to diverge from the correct value.

Note that this approach differs from what one might expect to be the method for calculating values for ln x. Most often, the natural logarithm is defined in terms of the definite integral

$$\ln x = \int_1^x \frac{1}{t}\, dt$$

TABLE 5.3

Evaluating ln(2) using limits

h	p
1.	1.
.0625	.708381
.0123458	.696121
.0039062	.694086
.0016	.693532
7.71605 E-4	.693332
4.16493 E-4	.693247
2.441406 E-4	.693207
1.52416 E-4	.693183
1.E-4	.693169
6.83013 E-5	.693160
4.82253 E-5	.693162

Based on what we learned previously about evaluating definite integrals, we might naturally attempt to use this definition, in conjunction with one of the numerical techniques of integration, to calculate ln x for any given x. However, such an approach is not particularly worthwhile. To obtain reasonably accurate results, we would have to use the program SIMP with a rather large value for n, which in turn takes a considerable amount of time to perform the calculations on most microcomputers. As a result, this approach is less feasible, compared to the speed involved in using a program such as LOGLIM above. In turn, LOGLIM is slower than some of the methods actually employed to calculate values for e^x, which are closely related to the technique used earlier in the program EXPX.

SECTION 5.2 PROBLEMS

1 The program EXPX can be modified to calculate values for e^{bx} by changing line 70 to read

```
70   LET A = A * B * X/(N + 1)
```

Use this to calculate values for e^{3x} for $x = 0, 2, 5$.

2 Change line 70 in EXPX to read

```
70   LET A = A * SQR(X)/(N + 1)
```

and use it to calculate values for $e^{\sqrt{x}}$ for $x = 0, 1, 2$.

5.3 INVERSE FUNCTIONS

Suppose that a function $y = f(x)$ is given on an interval $[a, b]$. If the function is monotonically increasing or decreasing, then we can be sure that the inverse function $x = f^{-1}(y)$ exists on either the interval $[f(a), f(b)]$ or the interval $[f(b), f(a)]$, depending on the relative sizes of $f(a)$ and $f(b)$. We also have theorems in calculus that tell us that the inverse function itself is monotonic, is continuous, and is even differentiable when the original function is differentiable. Unfortunately, all this information does not necessarily mean that we can actually find the inverse function in closed form. In fact, for all but the simplest functions, $f(x)$, it is impossible to do so, even though the inverse is known to exist. See Figure 5.3 for the graphs of a function and its inverse.

The computer, however, does provide us with a tool for investigating the inverse function and for calculating some of its values. The problem is as follows: For a given value of y, can we find the corresponding value for the inverse function $f^{-1}(y)$? This is equivalent to asking: for a given value y, can we calculate the value of x so that $f(x) = y$? In turn, this is equivalent to locating a root of the equation

$$g(x) = f(x) - y = 0$$

FIGURE 5.3

Graph of eˣ (solid curve) versus its inverse ln x (dashed curve)

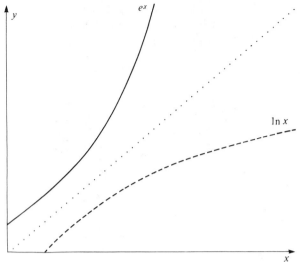

Note: x ranges from 0 to 7.65494 in steps of .765494
y ranges from 0 to 7.65494 in steps of .765494

since y has been specified. This type of problem can be solved, usually by the Bisection Method and the program BISECT or by Newton's Method and the program NEWTON. In either case, the program is applied to the function $f(x) - y$, where the value of y has to be supplied first. We will employ the Newton's Method approach here.

The following program is designed to carry out the required calculations. Since the quantity x to be found must lie within the original interval $[a, b]$, the program is designed to use the midpoint of the interval as the initial estimate x_0 for Newton's Method, $x_0 = (a + b)/2$. Further, the program is designed to accept the original function $f(x)$ at line 40 rather than $g(x) = f(x) - y$. In addition, we have to provide the first derivative of $f(x)$ at line 50.

Program FINVRS

```
10   REM INVERSE FUNCTIONS
     - FINDING X FOR GIVEN Y
20   INPUT "WHAT IS THE INTERVAL A,B? ";A,B
30   INPUT "WHAT IS THE VALUE OF Y? ";Y
40   DEF FNY(X) = ...
50   DEF FND(X) = ...
60   DEF FNG(X) = FNY(X) - Y
70   LET X0 = (A + B)/2
80   LET X1 = X0 - FNG(X0)/FND(X0)
90   PRINT X1,FNG(X0)
100   IF ABS(X1 - X0) < .0001 THEN 140
110   LET X0 = X1
120   GO TO 80
140   PRINT "CORRESPONDING TO Y, X = ";X1
150   END
```

When using this program, several important points must be kept in mind. First, the function $f(x)$ used must be monotonic on the indicated interval $[a, b]$. Second, the value for y used must be in the range of the original function $f(x)$. For instance, if $f(x) = \sqrt{x}$ on $[1, 10]$, then we could not use $y = 5$ (which corresponds to $x = 25$), let alone $y = -5$.

EXAMPLE 5.9

Consider the function

$$f(x) = \sqrt{e^x + x}$$

We note that for any value of $y \geq 0$, it is literally impossible to solve the equation $\sqrt{e^x + x} = y$ for x in closed form. If we apply the program FINVRS, however, we can obtain such values. The first derivative of $f(x)$ is

$$\left(\frac{1}{2}\right) \frac{e^x + 1}{\sqrt{e^x + x}}$$

which is always positive, so that the function is monotonically increasing and hence the inverse function exists. Suppose we use the interval $[0, 4]$ for x and select $y = 5$. This is a legitimate choice since $f(0) = 1$ and

$$f(4) = \sqrt{e^4 + 4} = 7.654943$$

The corresponding output from the program FINVRS is

x	$g(x)$
3.414158	.814252
3.112530	.058693
3.087237	3.684517E-4
3.087076	1.676381E-8
CORRESPONDING TO Y, X = 3.087076	

This can easily be checked, and we indeed find that $f(3.087076) = 4.99999869$.

EXERCISE 4

Modify the program FINVRS to provide a test for the derivative being close to 0.

EXERCISE 5

Modify FINVRS to check whether the successive terms calculated begin to diverge due to round-off errors.

EXERCISE 6

Since Newton's Method may diverge if the initial estimate is too far from the root, we may have problems taking the midpoint of the interval $[a, b]$ if the interval is quite large. To circumvent this possible problem, modify FINVRS so that the Bisection Method is applied first to close in on the root, say until the endpoints of the subinterval are within .01 of each other, and then use Newton's Method to zero in on the root quickly.

If we apply the program FINVRS or one of the modifications of it using a succession of different values of y to obtain the corresponding set of values for x, then we will have effectively generated a table of values for the inverse function $f^{-1}(y)$. We then can plot these ordered pairs (x, y), connect them, and so obtain the graph of the inverse function.

EXERCISE 7   Modify FINVRS to provide for a loop that will calculate the x values corresponding to each member of a sequence of y values.

SECTION 5.3 PROBLEMS

Use FINVRS or one of your modifications of it to calculate a sequence of at least 10 points on the curve for the inverse of the following functions. Plot these points and sketch a portion of the graph.

1 $f(x) = x^3$ for x from -1 to 4

2 $f(x) = \sin x$ on $\left[-\dfrac{\pi}{4}, \dfrac{\pi}{4} \right]$

3 $f(x) = e^x + x$ on $[0, 4]$

4 $f(x) = \sqrt[3]{2x - 4}$ on $[2, 6]$

5 $f(x) = \dfrac{20}{.05 + 1.95e^{-2x}}$ on $[0, 3]$

5.4
GROWTH AND DECAY PROBLEMS: UNINHIBITED AND INHIBITED MODELS

One of the most common situations where calculus is applied is in growth and decay problems. In these cases, we deal with some quantity a that is changing in such a way that the rate of change, da/dt, is proportional to the amount of a at each instant of time. That is,

$$\frac{da}{dt} = ka$$

and the corresponding solution is given by

$$a = a_0 e^{kt}$$

where a_0 is the initial amount of a present and k is the constant of proportionality. If the time is measured from some initial value t_0 other than 0, then the solution is

$$a = a_0 e^{k(t - t_0)}$$

When the constant of proportionality $k > 0$, the quantity a is growing; when $k < 0$, the quantity a is decaying. Many processes in science, business, and the social sciences — such as radioactive decay, dilution of solutions, population growth, interest with continuous compounding, learning models, and depletion of natural resources — all appear to fit this mathematical model. Each of these is considered an example of uninhibited growth or decay.

The following program, GRWDCY, is a simple way of using the computer to calculate the amount of a for any length of time, based on the growth and decay model.

Program GRWDCY

```
10   REM PROGRAM FOR UNINHIBITED GROWTH AND
     DECAY
20   INPUT "WHAT IS THE CONSTANT K? ";K
30   INPUT "WHAT IS THE INITIAL AMOUNT OF A?
     "; A0
40   INPUT "HOW LONG SHOULD THE PROJECTION
     RUN? ";T2
50   FOR T = 0 TO T2
60   PRINT T, A0 * EXP(K * T)
70   NEXT T
80   END
```

EXAMPLE 5.10

Suppose that $1000 is deposited in an account paying 12% per year compounded continuously. (The difference between continuous compounding and daily compounding is relatively slight over the short run, but the mathematics involved is quite different.) The growth constant k is .12, and so, over the first 10 years, the program yields

```
0        1000.00
1        1127.50
2        1271.25
3        1433.33
4        1616.07
5        1822.12
6        2054.43
7        2316.37
8        2611.70
9        2944.68
10       3320.12
```

EXERCISE 8

Modify GRWDCY to provide for appropriate headings and for the use of an initial time t_0. Also provide for a step to be used, if desired, so that only a relatively few numbers will be printed out. Finally, make sure that the length of the projection, t_2, is greater than the initial time, t_0.

As it stands, the program GRWDCY cannot be used directly to handle the typical problems in radioactive decay, since the constant k usually is not given explicitly. Rather, we might be told that the half-life of radium, say, is 1580 years with an initial quantity a_0 present. From this, we could calculate the constant k as follows: when $t = 1580$, $a = a_0 e^{k \cdot 1580} = a_0/2$, so that

$$e^{1580k} = .5$$

and so

$$\ln e^{1580k} = 1580k = \ln .5$$

From this, we find

$$k = \frac{\ln .5}{1580} = -.0004387$$

With this value, we can now apply the program. However, there is one problem with applying the program as it appears above. The resulting output will list the quantity of radium each year starting from $t = 0$ and will require 1580 lines just to reach the half-life point at $t = 1580$. For most such problems, this would be totally unnecessary. The modification to the program suggested in Exercise 8 above provides for introducing a step, say 100 years, that will produce only the value every 100 years, and so cut down the output to a manageable 16 lines.

EXERCISE 9

Modify the program GRWDCY to accept the following data: the initial quantity a_0 at time t_0 and the final quantity a_1 at some later time t_1. Use this data in the program to calculate the value for the constant k. Note that the functional expression for a given above must be changed to account for the presence of the initial time t_0.

The uninhibited growth and decay models just discussed suffer from one serious drawback. In most of the situations where the models are applied, it is extremely unrealistic to pursue the process to the limit along an exponential growth or decay curve. These processes rarely continue forever. For example, population growth might appear to grow exponentially, but this growth cannot go on indefinitely — eventually, the mathematics would predict more people on earth than there are atoms in the universe. Obviously, then, the exponential growth model applies only for a limited time — thereafter, other factors, such as famine, plague, war, or emigration, come into play to inhibit or curtail further population growth. A mathematical model for such inhibited growth or decay is given by a rate of growth for the quantity a given by

$$\frac{da}{dt} = ka - ma^2$$

Typically, the constant m is much smaller than k. As a result, while t and therefore a are relatively small, the term $-ma^2$ has little effect and the above equation is almost $da/dt = ka$ to produce exponential growth. However, as t and so a increase, then the value for da/dt begins to decrease, and so the slope of the growth curve begins to level out. The corresponding curve is shown in Figure 5.4 and is known as a *logistic* or inhibited growth curve.

To obtain a formula for this curve, suppose that the initial quantity of a at time t_0 is a_0. Therefore, we find that

$$\frac{da}{ka - ma^2} = dt$$

so that, upon integrating both sides from t_0 to t, we find

$$\int_{t_0}^{t} \frac{da}{ka - ma^2} = \int_{t_0}^{t} dt$$

We evaluate this integral using a partial-fractions expansion. Thus,

$$\frac{1}{a(k - ma)} = \frac{p}{a} + \frac{q}{k - ma}$$

from which we find

$$p = \frac{1}{k} \quad \text{and} \quad q = \frac{m}{k}$$

Therefore, considering the indefinite integrals, we obtain

$$\int \frac{da}{a(k - ma)} = (1/k) \ln a + (m/k) \ln(k - ma)(-1/m)$$

$$= (1/k)[\ln a - \ln(k - ma)]$$

$$= (1/k) \ln \frac{a}{k - ma}$$

$$= \int dt = t + C_1$$

or equivalently,

$$\ln \frac{a}{k - ma} = kt + kc_1 = kt + C_2$$

Taking exponentials of both sides, we find that

$$\frac{a}{k - ma} = e^{kt + C_2} = Ce^{kt}$$

If we now use the fact that, at $t = t_0$, the initial quantity present is a_0, then

FIGURE 5.4

Graph of logistic growth curve
(solid) versus exponential
growth curve (dotted)

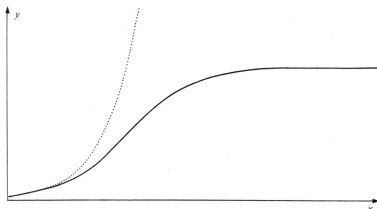

Note: x ranges from 0 to 10 in steps of 1
y ranges from 0 to 3000 in steps of 300

we obtain

$$\frac{a_0}{k - ma_0} = Ce^{kt_0}$$

from which

$$C = \frac{a_0}{k - ma_0} e^{-kt_0}$$

and so

$$\frac{a}{k - ma} = \left(\frac{a_0}{k - ma_0} e^{-kt_0}\right)(e^{kt}) = \frac{a_0}{k - ma_0} e^{k(t-t_0)}$$

If we multiply out this last equation and collect the terms involving a, then we finally obtain the expression for the general solution of the logistic equation:

$$a(t) = \frac{ka_0}{ma_0 + (k - ma_0)e^{-k(t-t_0)}}$$

We note that $m = 0$ in this formula causes it to reduce simply to the usual result for exponential growth or decay, $a = a_0 e^{k(t-t_0)}$. Further, as $t \to \infty$, the exponential term in the denominator approaches zero, and so the entire expression approaches a limit of k/m. This ratio is the limiting value for a in the sense that it is a horizontal asymptote. In terms of population, it is the maximum population that a region can support. Moreover, for a given pair of values k and m, the function will always approach this limiting ratio k/m regardless of the initial quantity a_0.

EXAMPLE 5.11   Suppose that 10 bacteria are introduced into a culture medium and that the population doubles every day, so that $k = 2$. Suppose that $m = .005$. The population at any time t will then be given by

$$a = \frac{(2)(10)}{(.005)(10) + [2 - (.005)(10)]e^{-2t}}$$

$$= \frac{20}{.05 + 1.95e^{-2t}}$$

If we modify line 60 of the program GRWDCY to include this expression instead of simply $a_0 e^{kt}$, then we obtain the output shown in Table 5.4. A plot of these points is shown in Figure 5.5, from which we see that they do fall onto a logistic curve. Moreover, from either table or the graph, it is clear that the limiting value for this population is 400, which is equal to the ratio $k/m = 2/.005 = 400$.

EXERCISE 10   Modify GRWDCY to accept appropriate input values and the formula for the logistic or inhibited growth model.

TABLE 5.4

Values for the Inhibited Growth Model for Bacteria

t	$a(t)$
0	10
1	63.71
2	233.33
3	364.74
4	394.83
5	399.29
6	399.90
7	399.99

FIGURE 5.5

Logistic growth curve for bacterial population

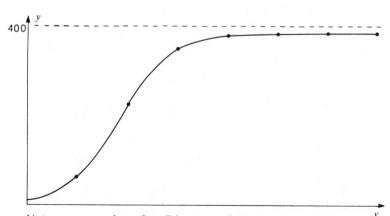

Note: x ranges from 0 to 7 in steps of .7
y ranges from 0 to 420 in steps of 42

SECTION 5.4 PROBLEMS

1 Suppose $1000 is deposited into an account paying 5.5% interest compounded continuously. Use program GRWDCY to generate a table showing the balance for the first 20 years.

2 Suppose the Indians who sold Manhattan Island in 1626 deposited their $24 in an account paying 6% compounded continuously. Use program GRWDCY to find the current value of their investment.

3 A blob in a horror movie increases its diameter by a factor of 1.7 every 24 hours (so $k = .7$). If it is first observed as being 2 feet across, how large is it after one week? After one month (30 days)? How long does it take until it covers the entire United States (3,000 miles)? (Hint: Use either program BISECT or NEWTON to obtain an accurate answer to the last part.)

4 Assume world population is growing exponentially at the rate of 2% per year. If the current population is 3 billion, find how long it will take for the population to hit 5 billion.

5 Suppose a dose of a certain drug is 600 milligrams. If the effectiveness decreases at a rate of 25% of the drug dose per hour, how long would it take to bring the drug level down to the equivalent of 100 mg? To under 5% of the original level?

6 Use the same data as in Example 5.11 to determine how long it will take for the bacterial population to reach 100.

7 Use the same data as in Example 5.11. Suppose the original count of the bacteria is 1000. Use the program TABLE to determine the population over the first 10 days. Can you account for the behavior of the population figures? Use this data to sketch a graph of the population function. What does the graph suggest?

5.5
THE TRIGONOMETRIC FUNCTIONS

In Section 5.1, we discussed how the computer can be applied to most of the types of calculus problems we previously encountered when the function involved consists of exponential and logarithmic terms. In a comparable way, we can deal with functions involving the trigonometric functions. Almost all versions of the BASIC language have built-in functions for some of the trigonometric functions. Minimally, the sine of any angle x, in radians, can be called on by simply typing SIN(X). Most versions also provide COS(X) and TAN(X). If these latter two functions are not available in a particular version of BASIC, it is fairly simple to obtain them using several of the standard trigonometric identities. For example, since $\sin^2 x + \cos^2 x = 1$, we have

$$\cos x = \pm\sqrt{1 - \sin^2 x}$$

which in BASIC would become

```
SQR(1 - SIN(X)^2)
```

Note that the exponent *follows* SIN(X) unlike the usual algebraic convention of writing $\sin^2 x$. If we try to write SIN^2(X), say, then the computer responds with an error message. Further, SIN(X^2) refers to the sine of x^2, and so on.

Once $\sin x$ and $\cos x$ are available, we can easily calculate the tangent, even if the TAN function is not available—by using $\tan x = \sin x/\cos x$. Similarly, even though $\cot x$, $\sec x$ and $\csc x$ are not built-in functions in BASIC, we can find them easily enough using the identities

$$\cot x = 1/\tan x = \cos x/\sin x$$

$$\sec x = 1/\cos x$$

$$\csc x = 1/\sin x$$

In addition, most versions of BASIC also provide a built-in function for the inverse tangent. It is usually denoted by ATN(X) for arc tangent. Thus, if

$$t = \tan a \qquad \text{then} \qquad a = \text{Tan}^{-1} t$$

Equivalently, in BASIC, if

```
T = TAN(A)    then    A = ATN(T)
```

where a is in radians and is in the open interval $(-\pi/2, \pi/2)$. Using this and appropriate identities, we can obtain formulas for the inverse sine and inverse cosine even though they are usually not part of the BASIC language. First, if

$$s = \sin a \qquad \text{then} \qquad a = \text{Sin}^{-1} s$$

where a is in the closed interval $[-\pi/2, \pi/2]$. Similarly, if

$$c = \cos a \qquad \text{then} \qquad a = \text{Cos}^{-1} c$$

where a is in the closed interval $[0, \pi]$.

Now, since $t = \tan a = s/c$, it follows that

$$a = \text{Tan}^{-1}\left(\frac{s}{c}\right) = \text{Tan}^{-1}\left(\frac{s}{\sqrt{1 - s^2}}\right)$$

Therefore, in BASIC, if we have to find the inverse sine of a number s, we would simply work with the expression

```
A = ATN(S/SQR(1 - S^2))
```

Similarly, to find the inverse cosine of a number c, if $c > 0$, we use

```
A = ATN(SQR(1 - C^2)/C)
```

If $c < 0$, then

```
A = ATN(SQR(1 - C^2)/C) + PI
```

It is essential to keep in mind that all trigonometric operations performed on the computer are done using radian measure. If you ever need to work with degrees, you must provide the conversion based on

$$\pi \text{ radians} = 3.14159 = 180°$$

as part of the program.

With these facts in hand, we are now able to apply all of the previous programs to functions involving the trigonometric functions.

EXAMPLE 5.12

Find all maxima and minima for the function $f(x) = x^2 \sin x$ on the interval $[-\pi, \pi]$. See Figure 5.6.

If we attempt this problem by hand, we would seek all critical points where $f'(x) = 2x \sin x + x^2 \cos x = 0$. One such point is clearly at $x = 0$; the others, if they even exist, cannot be found in closed form. However, if we first apply the program INCDEC from Section 2.2 with step $h = .1$ to the given function $f(x)$, then we obtain

$f(x)$ decreases from $-\pi$ to -2.3,

$f(x)$ increases from -2.3 to 2.3,

$f(x)$ decreases from 2.3 to π.

We next use the values -2.3 and 2.3 as input to the program NEWTON from Section 4.5, applied to the derivative function $f'(x)$. (Remember, we seek the roots of $f'(x)$, not $f(x)$.) It is not necessary to use $x = 0$, since it is clear that it is the precise value for one of the roots. NEWTON then pro-

FIGURE 5.6

Graph of f(x) = x² sin x on
[−π, π]

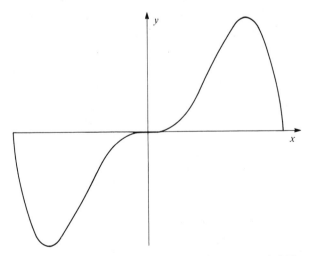

Note: x ranges from -3.14 to 3.14 in steps of .628
y ranges from -3.94526 to 3.94526 in steps of .789053

duces critical points at -2.28892973 and at 2.28892973. Finally, we supply the three critical points as input to the program MAXMIN from Section 3.6 and obtain the appropriate conclusions:

There is a minimum at $x = -2.28892973$

There is neither a maximum nor a minimum at $x = 0$

There is a maximum at $x = 2.28892973$.

EXAMPLE 5.13   Find the length of arc of the curve given by $y = \sec x$ from $x = 0$ to $x = \pi/4$.

Since $f'(x) = \sec x \tan x$, the arc length is given by

$$s = \int_0^{\pi/4} \sqrt{1 + \sec^2 x \tan^2 x}\, dx$$

Unfortunately, this integral cannot be evaluated in closed form, and so we must resort to a numerical method to approximate its value. We use the program SIMP from Section 4.3 applied to the function

```
FNY(X) = SQR(1 + SIN(X)^2/COS(X)^4)
```

with $n = 500$ subdivisions to obtain a value of .924674 unit as the arc length.

EXAMPLE 5.14   Find all points of intersection of the curves given by $f_1(x) = \sin 2x$ and $f_2(x) = 2 \cos x$ between $x = 0$ and $x = 5$. See Figure 5.7.

The points of intersection are equivalent to the roots of the function $g(x) = \sin 2x - 2 \cos x = 0$. These cannot be found in closed form; there-

FIGURE 5.7

Graphs of $f_1(x) = \sin 2x$ (solid)
and $f_2(x) = 2 \cos x$ (dotted)

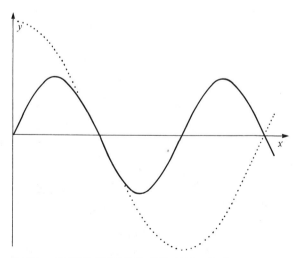

Note: x ranges from 0 to 5 in steps of .5
y ranges from -1.99993 to 2 in steps of .399993

fore, we apply Newton's Method to the problem via the program NEWTON. If we use the program with a variety of initial values for x_0 on the interval $[0, 5]$, then we obtain convergence to one of two roots, either $x = 1.569895$ or $x = 4.712389$.

It is worth noting that the function used here has a great many other roots, though no others are inside the indicated interval. This is due to the periodic nature of the functions involved. If we apply NEWTON with any initial value from 0 to approximately 3.6, then Newton's Method invariably converges to the first root. On the other hand, if we start with any initial estimate from about 3.8 to 5, then the method invariably converges to the second root. However, the graph of the function is extremely flat between 3.6 and 3.8 (the derivative $g'(x)$ has a root near $x = 3.66519$), so the tangent line to the curve at any such point shoots far outside the indicated interval. As a result, if we choose an initial value between 3.6 and 3.8, the algorithm converges to some other root of $g(x)$ outside the interval.

SECTION 5.5 PROBLEMS

Apply any of the programs from the previous chapters that are necessary to solve the following problems.

1 Find the following limits, if they exist.

(a) $\displaystyle\lim_{x \to \pi/2^+} \frac{\tan x}{\tan 3x}$

(b) $\displaystyle\lim_{x \to 0} \left(\frac{1 + 3x}{\sin x} - \frac{1}{x} \right)$

(c) $\displaystyle\lim_{x \to \pi} \frac{5 \sin^2 x}{1 + \cos x}$

(d) $\displaystyle\lim_{x \to 0} \frac{x - \text{Tan}^{-1} x}{x - \text{Sin}^{-1} x}$

(e) $\displaystyle\lim_{x \to 0} \left(\frac{1}{x} - \text{Sin}^{-1} x \right)$

2 Find the point of intersection of $f_1(x) = x \, \text{Tan}^{-1} x$ and $f_2(x) = x - 3 \cos x$.

3 Find all relative maxima and minima for the following functions:

(a) $f(x) = (1 - \cos x)^x$ on $[0, 2\pi]$

(b) $f(x) = 6x^{(1 - \cos x)}$ on $[0, 2\pi]$

4 Find the area under the curve $f(x) = x \sec x$ from $x = 0$ to $x = \pi/4$.

5 Find the arc length for the curve in Problem 4.

6 Find the volume of the solid of revolution obtained when the region in Problem 4 is rotated about the x-axis; about the y-axis.

5.6
THE HYPERBOLIC FUNCTIONS

In the previous section, we saw how the trigonometric functions could be handled by the computer. When SIN(X), COS(X), and TAN(X) are available as part of the BASIC language, their use is simple. Even when we have to deal with the other trigonometric functions, it is just a matter of expressing them in terms of the known quantities.

A similar situation holds for the hyperbolic functions. There are no built-in functions for them in the BASIC language, but they can be expressed easily in terms of available functions; in fact, that is precisely how they are defined. We have

$$\sinh x = \frac{e^x - e^{-x}}{2}$$

$$\cosh x = \frac{e^x + e^{-x}}{2}$$

$$\tanh x = \frac{\sinh x}{\cosh x} = \frac{e^x - e^{-x}}{e^x + e^{-x}}$$

and so forth. As a result, it is now possible to apply all of the previous programs to functions involving the hyperbolic functions.

EXAMPLE 5.15   Find a number c satisfying the Mean Value Theorem for the function $f(x) = \sinh x + (x + 1) \cosh x$ on the interval $[-2, 3]$.

If we apply the program MVT from Section 3.5 to the function

```
FNY(X) = (EXP(X) - EXP(-X))/2 + (X + 1)
         * (EXP(X) + EXP(-X))/2
```

then we obtain the value $c = 1.61463219$. See Figure 5.8.

FIGURE 5.8

Mean Value Theorem applied to $f(x) = \sinh x + (x + 1) \cosh x$ on $[-2, 3]$

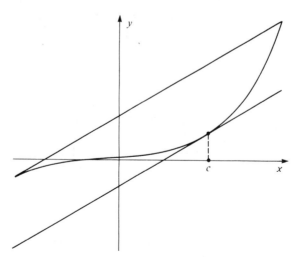

Note: x ranges from -2 to 3 in steps of .5
y ranges from -32.2136 to 50.2885 in steps of 4.34222
$c = 1.61463$

FIGURE 5.9
Solid of revolution produced by
$f(x) = \ln x \tanh x$ on $[1, 2]$

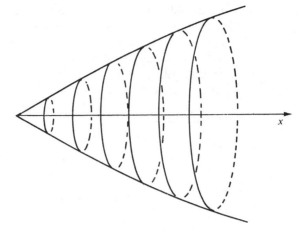

Note: x ranges from 1 to 2.13364 in steps of .113364
y ranges from − .668213 to .668213 in steps of .133643
The volume of the solid of revolution is .552015

EXAMPLE 5.16 Find the volume generated when the region under the curve given by $f(x) = \ln x \tanh x$ between $x = 1$ and $x = 2$ is rotated about the x-axis.
The indicated volume is given by

$$V = \int_1^2 \pi (\ln x)^2 \tanh^2 x \, dx$$

and cannot be evaluated in closed form. We therefore apply the program SIMP from Section 4.3 with $n = 500$ subdivisions to the function

```
FNY(X) = 3.14159 * LOG(X)^2 * ((EXP(X)
         - EXP(-X))/(EXP(X) + EXP(-X)))^2
```

and so obtain a value of .552015 cubic unit for the volume. See Figure 5.9.

SECTION 5.6 PROBLEMS

Apply any of the programs from the preceding chapters that are necessary to solve the following problems.

1 Find the following limits, if they exist.

(a) $\displaystyle\lim_{x \to 0} \frac{\sinh x}{x}$

(b) $\displaystyle\lim_{x \to 0} \frac{1 - \cosh x}{x}$

(c) $\displaystyle\lim_{x \to \infty} \frac{\sinh x}{e^x}$

2 Find the point on the curve given by $f(x) = \cosh x$ that is closest to $P(1, 0)$. (Hint: Consider the square of the distance.)

3 Find the equations of the two tangent lines at the point of intersection for the two curves $f(x) = \sinh x$ and $g(x) = 2 - \cosh x$ between $x = 0$ and $x = 1$.

4 Find the area under the curve $f(x) = \tanh x$ from $x = 0$ to $x = 1$.

5 Find the volume of the solid of revolution obtained by rotating the region in Problem 4 about the x-axis; about the y-axis.

5.7
COMPUTER SOLUTIONS OF DIFFERENTIAL EQUATIONS

A *differential equation* is an equation involving the derivatives of a function $y = f(x)$ with respect to the variable x. For example,

(1) $\quad y' - 3y = x^2$

(2) $\quad xy' + y \sin xy = e^x$

(3) $\quad y'' - 5y' + 6y = 4 \cos x$

Differential equations that involve only the first derivative, such as (1) and (2) above, are called first order differential equations. Those that involve derivatives up to the second, such as (3), are second order differential equations, and so forth. The solution of a differential equation is that function $y = f(x)$ that satisfies the equation. Thus, for (1) above, the general solution is given by

$$y = Ce^{3x} - \frac{x^2}{3} - \frac{2x}{9} - \frac{2}{27}$$

while for (3), the general solution is

$$y = Ae^{2x} + Be^{3x} + \frac{2}{5} \cos x - \frac{2}{5} \sin x$$

That these are the solutions can be verified easily by simply substituting the given expressions into the respective differential equations. The quantities A, B, and C denote arbitrary constants analogous to the constants of integration we have seen before. Also, note that the general solution to (1), a first order differential equation, involved one such constant, while the solution to (3) required two of them. Further, every possible value for these constants will give rise to a different particular solution to the differential equation.

The other thing that you may have noticed is that we have not given a general solution to equation (2). In fact, while some powerful theorems in differential equations assure us that this equation does have a solution, there are no known methods to actually *find* that solution in closed form. Unfortunately, the overwhelming majority of the differential equations that arise in practice (as opposed to those that are treated and solved in courses in differential equations) fall into the same category as (2). They cannot be solved explicitly, and, as a result, the methods usually used in real-world situations invariably involve approximate numerical solutions generated by computers.

In the present section, we consider only first order differential equations that can be written in the form

$$y' = f(x, y)$$

and ignore more complicated cases where there is an implied relationship involving x, y, and y', as in

$$y \tan(xy') = \frac{\sqrt{y}}{x} + y' e^{(x/y)} + 5$$

With this in mind, equation (2) would be rewritten as

$$y' = f(x, y) = \frac{e^x - y \sin xy}{x} \qquad x \neq 0$$

Since the object of solving a differential equation is to find y by itself (when possible) or, equivalently, to eliminate the derivative y', a general procedure would involve some type of integration. Thus, for the particularly simple differential equation

$$y' = 3x^2 + 4x - 7e^{5x}$$

a single integration yields

$$y = x^3 + 2x^2 - \frac{7e^{5x}}{5} + C$$

where C is a constant of integration. Similarly, the solution of a second order differential equation would seemingly involve two integrations and so would produce two arbitrary constants, and so on.

Consider the differential equation $y' = f(x, y)$ and suppose that we are interested only in solutions starting at some initial value of x, say x_0. This is often thought of as the initial time for the solution. At this initial instant x_0, there are many possible initial values for the function y — in fact, any value y_0 will produce a different solution. Each solution can be thought of geometrically as a curve starting at time x_0 with initial height y_0 and extending to the right, as shown in Figure 5.10. This produces a family of solutions to the differential equation, one for each initial point (x_0, y_0), or equivalently, one corresponding to each possible value of the arbitrary constant of integration. In fact, selection of a particular set of initial conditions (x_0, y_0) is used to determine a specific value for the constant and so determines the unique solution through the given initial point.

Unfortunately, it is usually impossible to perform the actual integration. Therefore, we consider instead the possibility of solving differential

FIGURE 5.10

Family of solution curves to a
differential equation

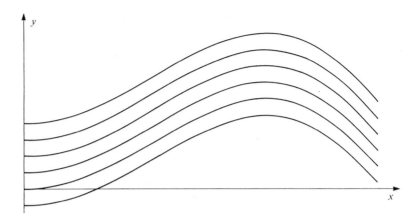

equations in a numerical manner. One thought that probably seems obvious
is to use

$$y' = \frac{dy}{dx} = f(x, y)$$

to write

$$y = \int_{x_0}^{b} f(x, y) \, dx = \int_{x_0}^{b} f(x, y(x)) \, dx$$

where x_0 is at the initial point for the solution and the upper limit b is the
rightmost value to which the solution will extend. Then it makes sense to
apply some numerical integration technique, such as Simpson's Rule, to
evaluate the integral. This procedure is not as easy as it sounds and, in fact,
is complicated by the fact that y is some unknown function of x inside the
integrand.

A more productive approach is to apply approximation methods to
the derivative y' instead of to the integral. The resulting technique is known
as Euler's Method. From the definition of derivative,

$$y' = \lim_{h \to 0} \frac{y(x + h) - y(x)}{h} = f(x, y(x))$$

Therefore, if h is small,

$$\frac{y(x + h) - y(x)}{h} \approx f(x, y(x))$$

so that

$$y(x + h) - y(x) \approx f(x, y(x))h$$

or

$$y(x + h) \approx y(x) + f(x, y(x))h$$

We use this approximation to find, numerically, the solution to the differential equation $y' = f(x, y)$ starting at the initial point (x_0, y_0). The approximation then produces, using $x = x_0$ and $y = y_0$,

$$y(x_0 + h) \approx y(x_0) + f(x_0, y(x_0))h$$
$$= y_0 + f(x_0, y_0)h = y_1$$

If h is small, then this new y_1 is close to the exact value $y(x_0 + h)$, as shown in Figure 5.11. (This diagram is similar to the usual one drawn to show the differential, and, in fact, the two concepts are closely related.) As a result, it makes sense to use y_1 as if it were a point on the solution curve rather than an approximation and extend the solution curve still further to the right with

$$y(x_1 + h) = y(x_0 + 2h) \approx y(x_1) + f(x_1, y(x_1))h$$
$$= y_1 + f(x_1, y_1)h = y_2$$

We can continue this process indefinitely, thus producing a sequence of points approximating the solution curve

$$(x_0, y_0), (x_1, y_1), \ldots, (x_n, y_n), \ldots$$

where the horizontal separation is the constant h.

This set of points can be thought of as all lying on a curve that lies close to the actual solution curve for the differential equation. More accurately, however, we are using the differential equation to give the slope y' of a tangent line to the true solution. Thus, geometrically, we are constructing a series of line segments that hopefully will remain close to the true solution curve, as shown in Figure 5.12.

The problem with this approach is that each successive approximation to a point on the true curve involves using a previous approximation. As a consequence, we are continually compounding errors, and the sequence of

FIGURE 5.11

Geometric representation of Euler's Method

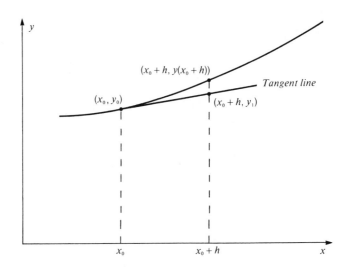

FIGURE 5.12

Series of line segment approximations via Euler's Method

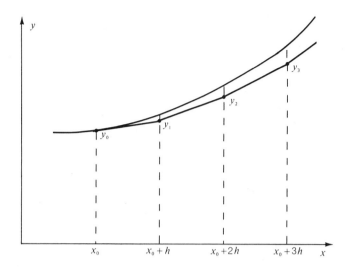

points will eventually diverge from the solution curve. This can be prevented temporarily by selecting a small value for h, but the error will still occur if the process is continued too long. Meanwhile, the small value for h translates into a large number of iterations and a corresponding increase in calculation time.

We illustrate Euler's Method in the following example.

EXAMPLE 5.17

Solve the differential equation $y' = f(x, y) = x^2$, subject to the initial conditions $x_0 = 0$ and $y_0 = 1$ (equivalently, $y(0) = 1$) from $x = 0$ to $x = 2$.

This differential equation can be integrated directly, so we obtain $y = x^3/3 + C$. Using the initial conditions, we quickly find that $C = 1$. Therefore, the actual solution is given by $y = 1 + x^3/3$, and we will use this to compare the accuracy of a variety of numerical solutions obtained using Euler's Method.

Suppose first we select $h = .5$. Therefore, we obtain

$$y_1 = y_0 + f(x_0, y_0)h = 1 + f(0, 1)(.5) = 1$$

compared to $y(.5) = 1 + (.5)^3/3 = 1.04167$. Similarly, we have

$$y_2 = y_1 + f(x_1, y_1)h = 1 + f(.5, 1)(.5) = 1.125$$

compared to $y(1) = 1.33333$. Similarly, $y_3 = 1.625$, compared to $y(1.5) = 2.125$ and $y_4 = 2.75$, compared to $y(2) = 3.66666$. Clearly, the approximation is not a good one and gets worse the further it is continued.

For comparison, suppose we take a smaller value for h, say $h = .1$, instead. We then obtain the results shown in Table 5.5. This table shows that the approximation is more accurate for the values of x near the initial point and that the level of accuracy remains higher for larger values of x. Nevertheless, the error terms do increase monotonically, and the approximation diverges if the solution is extended far enough. The accuracy can be in-

TABLE 5.5

Table of values illustrating Euler's Method

x	Euler	Actual	Error
0	1	1	0
.1	1	1.00033	-3.33309E-04
.2	1.001	1.00267	-1.66667E-03
.3	1.005	1.009	-3.99995E-03
.4	1.014	1.02133	-7.3334E-03
.5	1.03	1.04167	-.0116667
.6	1.055	1.072	-.0170001
.7	1.091	1.11433	-.0233334
.8	1.14	1.17067	-.0306667
.9	1.204	1.243	-.039
1	1.285	1.33333	-.0483335
1.1	1.385	1.44367	-.0586668
1.2	1.506	1.576	-.0700003
1.3	1.65	1.73233	-.0823336
1.4	1.819	1.91467	-.095667
1.5	2.015	2.125	-.110001
1.6	2.24	2.36533	-.125334
1.7	2.496	2.63767	-.141667
1.8	2.785	2.944	-.159001
1.9	3.109	3.28633	-.177334
2	3.47	3.66667	-.196667

creased by taking a still smaller value for h with the consequent increase in computation. See Figure 5.13, where a succession of values of h produces a series of piecewise linear approximations to the actual solution curve.

The above table, in part, was generated by the following program. Note that the columns for the actual values and the errors are not included, usually because the actual solution is not known. If it were, there would be no point in using this or any other approximation procedure.

FIGURE 5.13

Euler approximations with $n = 4, 8, 16, 32, 64,$ and 128 subdivisions

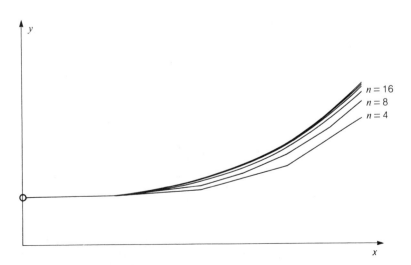

Program EULER

```
10   REM EULER'S METHOD TO SOLVE DIFFERENTIAL
     EQUATIONS
20   DEF FNA(X,Y) = ...
30   INPUT "THE INITIAL CONDITIONS X0,Y0
     ARE ";X0,Y0
40   INPUT "HOW FAR SHOULD THE SOLUTION
     EXTEND? ";B
50   INPUT "WHAT IS THE STEP H? ";H
60   FOR X = X0 TO B STEP H
70   PRINT X,Y0
80   Y0 = Y0 + FNA(X,Y0) * H
90   NEXT X
100  END
```

EXAMPLE 5.18

Apply the program EULER to the differential equation (2) at the beginning of this section.

We set $FNA(X, Y) = (EXP(X) - Y * SIN(X * Y))/X$ and select as initial values $x_0 = .1$ (to avoid $x = 0$) and $y_0 = 1$. Using $h = .1$, we obtain the data in Table 5.6. Of course, not knowing the actual solution, we do not know how accurate any of these approximations are.

TABLE 5.6

Euler's Method applied to differential equation (2)

x	Euler
.1	1
.2	2.00534
.3	2.2246
.4	2.21559
.5	2.15945
.6	2.10834
.7	2.07694
.8	2.06994
.9	2.09033
1	2.14246

SECTION 5.7 PROBLEMS

Apply the program EULER to each of the following differential equations with the given initial values for x_0 and y_0, length of extension b, and indicated number of steps to determine the step length h. Wherever possible, compare the results from the program with the actual results obtained by integration.

1 $y' = 5, y_0 = 20, x_0 = 0, b = 10, n = 40$

2 $y' = 4x + 3, y(0) = 2, b = 5, n = 20$

3 $y' = 6x^2 - 2x + 7, y(1) = 4, b = 3, n = 12$

4 $y' = 6x^2 - 2x + 7, y(1) = 4, b = 3, n = 20$

5 $y' = 3y + x^2, y(0) = 6, b = 2, n = 30$

6 $y' = x \sin y, y(0) = \pi, b = 3, n = 36$

7 $y' = x \sin(x + y), y(0) = 1, b = 3, n = 36$

8 $y' = x \sin(x + y), y(0) = 2, b = 3, n = 36$

9 $y' = x \sin(x + y), y(0) = 3, b = 3, n = 36$

10 Use every third point obtained in Problems 7, 8, and 9 to plot the three different solutions to the differential equation on the same graph. What do you notice about the behavior of these solutions?

11 $y' = (x^2 + y)e^{-y}, y(1) = 0, b = 4, n = 30$

12 $y' = (x^2 + y)e^{-y}, y(1) = 3, b = 4, n = 30$

SIX
POLAR COORDINATES
AND PARAMETRIC EQUATIONS

6.1
POLAR COORDINATES

We now turn our attention to the subject of polar coordinates to see how the computer can be an extremely useful tool for solving many of the problems that typically arise.

The polar coordinates of a point $P(r, \theta)$ represent a way of locating the point at a distance r from the pole or origin and at an angle of θ measured in the positive direction from a horizontal polar axis. In most situations encountered, we think of the distance r as being a function of the angle θ; thus, $r = f(\theta)$. In all the programs we have used until now, the independent variable has always been x. The computer has no objection to the use of any other letter as the variable; however, we cannot make use of the Greek letter θ since the computer has no key for it. As a result, we will use the letter Q as the variable, since it looks the most like θ. Thus, we will consider functions of the form $r = f(Q)$ when dealing with polar coordinates.

In any introductory study of polar coordinates, one of the standard problems is to sketch the graph of a polar curve. This is usually done by calculating a table of values for the function, plotting the resulting points, and connecting them with a smooth curve. The most time-consuming part of this procedure is obtaining the entries in the table, and this chore can be done very simply by the computer. It is just a matter of using a modification of the program TABLE from Section 2.1.

Before proceeding to such a program, we must keep in mind the fact that the computer deals with angles measured in radians. On the other hand, most polar graphing is done using the standard angles in degrees, namely 30°, 45°, 60°, and so on. Therefore, it is necessary to convert from radians to degrees within the program. Since π radians = 180°, it follows that $1° = \pi/180$ radians. Further, we will define all polar functions in the form FNR(Q).

Program POLAR

```
10   REM PROGRAM TO GENERATE A TABLE OF POLAR
     VALUES
20   DEF FNR(Q) = ...
30   FOR Q = 0 TO 360 STEP 15
40   PRINT Q,FNR(Q * 3.14159/180)
50   NEXT Q
60   END
```

137

EXAMPLE 6.1 Consider the function $r = \cos^2 Q + \sin^4 Q$ and sketch its graph.

While the calculation of the values for this function can be carried out by hand with a calculator, it is much easier to make use of the program POLAR. When we apply the program with

$$\text{FNR(Q)} = \text{COS(Q)}^2 + \text{SIN(Q)}^4$$

we obtain the values found in Table 6.1. With these values available, it is a simple task to graph the corresponding curve, as shown in Figure 6.1.

When we use the program POLAR, several things have to be kept in mind. For one, the program assumes that Q ranges from 0 to 2π to complete a full curve. This certainly need not be the case. If a function involves a double angle, say $\cos 2Q$, then all that need be done is to go from 0 to π. Alternatively, if there is a half-angle, say $\sin Q/2$, then a full curve is traversed between 0 and 4π. Furthermore, the use of the step of 15 at line 30 of the program may not always be appropriate—it may provide far too

TABLE 6.1

Table of values for $r = \cos^2 Q + \sin^4 Q$

Q	r
0	1
15	.9375
30	.8125
45	.75
60	.8125
75	.9375
90	1
105	.9375
120	.8125
⋮	⋮

FIGURE 6.1

Graph of $r = \cos^2 Q + \sin^4 Q$

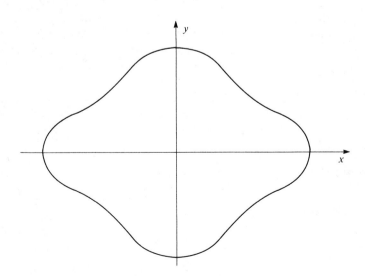

much information in some instances and not enough in others. Therefore, the step size might have to be altered for some functions.

Modify POLAR to provide appropriate headings and the use of INPUT statements to supply the desired interval and the step size for the values of the angle Q.

A second caution in using the program involves the problems that can arise with a curve such as the lemniscate where

$$r^2 = a \cos 2Q$$

for instance. It is very tempting to solve this expression for r,

$$r = \pm\sqrt{a \cos 2Q}$$

and define the function FNR(Q) accordingly for each sign. Unfortunately, this can create major problems if not done very carefully. Whenever the cosine becomes negative for some values of Q, we obtain a series of negative values for r^2; no corresponding points are plotted for such angles. Simply, the function is not defined for such values of Q. If this is anticipated in advance and a careful choice is made for the intervals of Q — for example, in the above case with $a > 0$

$$\left[-\frac{\pi}{4}, \frac{\pi}{4}\right] \quad \text{and} \quad \left[\frac{3\pi}{4}, \frac{5\pi}{4}\right]$$

then the program will work well. Alternatively, we may just define FNR(Q) = A * COS(2 * Q) to produce values for r^2, with the understanding that we then have to take all allowable square roots separately. We now turn to another standard type of problem in polar coordinates for which the computer can play a valuable role. We consider the problem of locating the points where two polar curves, $r = f(Q)$ and $r = g(Q)$, intersect. This can be handled easily by considering the difference of the two functions, $f(Q) - g(Q)$, and then applying either Newton's Method or the Bisection Method to the new function. We note that we can apply either program NEWTON or BISECT, with only minor modifications to reflect the fact that the variable is now being denoted by Q instead of by x. Actually, we could still work with the original versions of the programs using x, since the computer does not care what the variable stands for. The problem is to keep track of the significance of the variable as an angle, not as the horizontal distance. It is safer to make the change to Q to emphasize the polar variable.

As we have noted before, Newton's Method is usually the preferred choice because it converges at a much faster rate than the Bisection Method does. However, if the function has a horizontal tangent within a given interval, Newton's Method often converges to a root outside of the interval. When dealing with polar functions, we almost always find ourselves facing expressions involving sines and cosines. As a consequence, Newton's Method should be used with great care. If speed is of the essence, it still might be a good idea to apply the Bisection Method first to obtain a *good* approximation to the desired root and then use Newton's Method to zero in on the root quickly.

Whichever method is used, the typical problem to be solved involves locating all points of intersection of two given curves. Therefore, we usually have to first employ some type of search method to approximate the roots. Such an approach can use the program POLAR, applied to the difference of the two functions, to locate approximately where signs change. This information then can be used as input to either program NEWTON or BISECT to locate the root accurately, as shown in Example 6.2.

EXAMPLE 6.2

Find all points of intersection of $r = f_1(Q) = \sin Q + \cos Q$ and $r = f_2(Q) = 1 + 3 \cos Q$ in the interval $[0, 2\pi]$.

We first consider the function $f_1(Q) - f_2(Q)$ and so construct

$$r = f(Q) = -2 \cos Q + \sin Q - 1$$

and seek all possible roots between 0 and 2π. If we first apply a modification of the program POLAR to calculate the values for the function over a range of angles given in radians, then we find that the function changes sign between $Q = 1.5$ and $Q = 1.6$ and again between $Q = 3.7$ and $Q = 3.8$, based on a step of .1 radian. If we then apply the program BISECT with an initial interval 1.5 to 1.6, we find that one root occurs at $Q = 1.5707962$. We note that this is essentially $\pi/2$, and we can easily check by hand that there is a point of intersection there. If we then apply the program using an initial interval from 3.7 to 3.8, we locate the second root at $Q = 3.7850937$.

It is important that you realize that the above procedure applies only to solving for points of intersection that correspond to the same value for the angle Q. However, it is possible for two curves in polar coordinates to pass through one another at a point where the r-coordinates agree, but the Q-coordinates differ. This is quite common at the pole, as is seen in Figure 6.2. In particular, the solid curve, representing $f_1(Q)$, passes through the pole at an angle of $Q = -\pi/4$ while the dotted curve, representing $f_2(Q)$, passes through the pole at angles of $Q = \text{Cos}^{-1}(-1/3)$ and $2\pi - \text{Cos}^{-1}(-1/3)$.

The best way to handle such an eventuality is to sketch the graphs of the two functions, either analytically or by using an appropriate computer program (such as the method used in Example 6.1) and looking to see whether any apparent intersections occur that were not produced by the above procedure. Since this situation happens most frequently at the pole, you should be especially alert to the possibility when both curves pass through it. In that case, simply substitute $r = 0$ into both expressions for r and see whether or not the corresponding angles agree.

FIGURE 6.2

Graphs of $r = \sin Q + \cos Q$ (solid) and $r = 1 + 3 \cos Q$ (dotted)

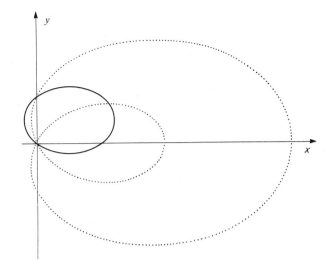

SECTION 6.1 PROBLEMS

Use the program POLAR to obtain a set of points that will enable you to sketch the graphs of the following polar curves.

1 $r = \cos^3 Q + \sin^5 Q$

2 $r = \cos Q + \cos^4 Q$

3 $r = 1 + \sin Q + \cos Q$

4 $r = (1 + \sin Q + \cos Q)^2$

5 $r = \sqrt{2 + \sin Q + \cos Q}$

Find all points of intersection in the interval $[0, 2\pi]$ of the following pairs of polar curves.

6 $r = \sin Q + \cos Q$ and $r = Q$

7 $r = 5 - 3 \cos Q$ and $r = 2 + \sin Q$

8 $r = 2(1 + \sin Q)$ and $r = 5(1 - \cos Q)$

9 $r = 4 \sin Q$ and $r = 3 + 2 \cos Q$

10 $r = \sin 2Q$ and $r = \cos 3Q$

6.2
AREA AND ARC LENGTH IN POLAR COORDINATES

The area of a region given in polar coordinates can be found from the integration formula:

$$\text{Area} = \int_\alpha^\beta (1/2) r^2 \, d\theta \quad \text{or} \quad \int_\alpha^\beta (1/2) r^2 \, dQ$$

When the relationship between r and Q is extremely simple, this integral can be evaluated in closed form using the standard integration techniques. However, it doesn't take much to make the expression for r^2 too complicated to permit us to integrate directly. As a result, we often have to resort to numerical methods such as Simpson's Rule to approximate the value of the integral. To implement this, we simply apply the program SIMP from Section 4.3 to the function $(1/2)r^2(Q)$. As we have seen before, we either can change the program to work with the new polar variable Q or use the program as it stands while interpreting x to represent the angle.

EXAMPLE 6.3

Find the area of the region bounded by the polar curve $r = f(Q) = (\sin Q)/Q$ and the lines $Q = \pi/6$ and $Q = \pi/4$.

We note that the area cannot be found exactly, since the integral

$$\int_{\pi/6}^{\pi/4} \frac{(1/2) \sin^2 Q}{Q^2} \, dQ$$

cannot be evaluated in closed form. If we therefore apply SIMP with $n = 500$ subdivisions, we obtain a value of .12117 square unit for the area. Alternatively, if we use $n = 1000$, then the program produces a value of .12141.

Suppose that we face the problem of finding the area of the region bounded between two polar curves, $r = f(Q)$ and $r = g(Q)$. The first step in solving this problem is to find the points of intersection of the two curves. This can be handled using either straight-forward algebraic methods, if the functions are simple enough, or the techniques described in the previous section. Once the points of intersection are known, they are usually used as the limits of integration in the formula

$$A = \int_{\alpha}^{\beta} \frac{1}{2} [g^2(Q) - f^2(Q)] \, dQ$$

EXAMPLE 6.4

Find the area of the region bounded between $r = f(Q) = 1 + 3 \cos Q$ and $r = g(Q) = \sin Q + \cos Q$.

In Example 6.2 of the last section, we saw that the points of intersection of these two curves occur at $Q = \pi/2$ and $Q = 3.7851$ (rounded to four decimal places). We therefore apply the program SIMP with $n = 1000$ to the function

```
FNY(X) = ((SIN(X) + COS(X))^2
         - (1 + 3 * COS(X))^2)/2
```

The result obtained is 1.58213 square units.

It is interesting to note that the answer to the above example, 1.58213, is surprisingly close to $\pi/2 = 1.5706$; we might, therefore, conjecture that the precise value is $\pi/2$. To test whether this is indeed the case, we must either perform the integration by hand to get a closed form solution (it can be done for this integrand) or obtain greater accuracy with the numerical methods. This latter approach involves more than just running the program SIMP with a larger value for n. In addition, we must provide more decimal places for the points of intersection. Despite this, the computer method cannot prove conclusively that the area is precisely $\pi/2$.

EXERCISE 2

Test the conjecture on the value of the area in Example 6.2 by using the program SIMP with greater accuracy.

One more point about the last example bears mentioning. We implicitly assumed that $f(Q) = 1 + 3 \cos Q < g(Q) = \sin Q + \cos Q$ when setting up the integral. Had we selected the functions the other way around, the answer would have turned out to be -1.58213. Obviously, we could go at this blindly and allow the sign of the answer to tell us the correct orientation of the curves after the fact, but that is a rather unsophisticated approach and can often be misleading. To determine the orientation in advance, all that we need do is to test the relative values of the two functions $f(Q)$ and $g(Q)$ at any one point for Q. This can be done by hand if there are any "nice" values for Q (say $Q = 0$ or π) or by the computer. Once $f(Q)$ and $g(Q)$ are defined (or even FNA(Q) = FNF(Q)^2 − FNG(Q)^2) and a value of Q is chosen in the interval, we merely type the command

    PRINT FNA(Q)

in immediate mode (outside a program with no line number), and the computer will respond with the corresponding value. From that, we can decide on the appropriate orientation. In a case where there are a variety of different points of intersection, then the same procedure would have to be applied between each pair of successive points in ascending order.

We now turn our attention to another topic involving integration in polar coordinates, namely finding the arc length of a polar curve. The corresponding formula is

$$ s = \int_{\alpha}^{\beta} \sqrt{r^2 + r'^2}\, dQ $$

where r' is the derivative of $r = f(Q)$ with respect to Q. It is highly unlikely that such an integral can be evaluated in closed form unless the function $r = f(Q)$ chosen is exceptionally simple. On the other hand, the methods we have at our disposal for approximating a definite integral allow us to find the arc length for almost any polar curve. All we need do is reinterpret the function used in the program SIMP, for example.

EXAMPLE 6.5

Find the length of arc of the boundary of the region bounded by $r = f(Q) = 1 + 3 \cos Q$ and $r = g(Q) = \sin Q + \cos Q$.

We have already seen that the two curves intersect at $Q = 1.5706$ and $Q = 3.7851$. The arc length in question is therefore the sum of the arc lengths for the two curves between the two values for Q. Using program SIMP with $n = 1000$ subdivisions applied to the first function, we obtain a value of 5.20278 units for the arc length of $f(Q)$. Similarly, the arc length for $g(Q)$ is found to be 3.12551 units. Therefore, the total arc length for the boundary of this region is approximately 8.32829 units.

SECTION 6.2 PROBLEMS

Find approximate values for the areas enclosed by the following polar functions on $[0, 2\pi]$.

1 $r = \cos^3 Q + \sin^5 Q$

2 $r = \cos Q + \cos^4 Q$

3 $r = 1 + \sin Q + \cos Q$

4 $r = (1 + \sin Q + \cos Q)^2$

Find approximate values for the areas enclosed between the following pairs of polar curves.

5 Inside $r = \sin Q + \cos Q$ and outside $r = Q$

6 Between $r = 5 - 3 \cos Q$ and $r = 2 + \sin Q$

7 Between $r = 4 \sin Q$ and $r = 3 + 2 \cos Q$

Find the arc lengths of the following polar curves between $Q = 0$ and $Q = \pi$.

8 $r = 1 + \sin Q + \cos Q$

9 $r = 4 + 4 \sin Q$

10 $r = 4 + 5 \sin Q$

6.3
PARAMETRIC EQUATIONS

We now turn to a study of parametric equations and how they can be used in conjunction with a computer. The parametric representation for a curve is the pair of equations

$$x = f(t)$$

$$y = g(t)$$

where t is the parameter having values on some interval $[a, b]$. For each value of t in this interval, the two equations determine a point $(x, y) = (f(t), g(t))$ in the plane. The totality of all such points form the curve in question.

In practice, if we were to graph such a parametric curve, we would need a set of points corresponding to a set of values of t, and we would connect them with a smooth curve. The following program allows us to find the desired points quickly and easily. Without such a program, the task of graphing most parametric curves becomes overwhelmingly tedious.

Program PARAMEQN

```
10   REM FINDING VALUES FOR PARAMETRIC
     EQUATIONS
20   INPUT "HOW MANY POINTS? ";N
30   INPUT "WHAT IS THE INTERVAL FOR T? ";A,B
40   DEF FNX(T) = ...
50   DEF FNY(T) = ...
60   FOR T = A TO B STEP (B - A)/N
70   PRINT T,FNX(T),FNY(T)
80   NEXT T
90   END
```

EXAMPLE 6.6

A hypocycloid is the path traversed by a fixed point on a circle of radius r_0 rolling around the inside rim of a larger circle of radius r_1. See Figure 6.3. The parametric equations of this curve are given by

$$x = (r_1 - r_0) \sin\left(\frac{r_0 t}{r_1}\right) - r_0 \sin\left(\frac{r_1 - r_0}{r_1} t\right)$$

$$y = (r_1 - r_0) \cos\left(\frac{r_0 t}{r_1}\right) + r_1 \cos\left(\frac{r_1 - r_0}{r_1} t\right)$$

We seek to sketch the graph of the hypocycloid where the inner rolling circle has radius $r_0 = 1$ and the larger outer circle has radius $r_1 = 3$. Using the program PARAMEQN with $n = 12$ points and the interval $[0, 6\pi]$, we obtain the set of values, rounded off to permit ease of graphing, found in Table 6.2. Before we actually graph these points, it is worth noting that there is a considerable degree of symmetry in the x-values and in the y-values. Moreover, the range of t from 0 to 6π was chosen intentionally

FIGURE 6.3

Describing a hypocycloid

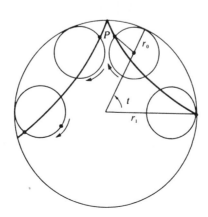

TABLE 6.2

Tables of values
for hypocycloid
with $r_0 = 1$ and $r_1 = 3$

t	x	y
0	0	3
1.57	.134	2.23
3.14	.866	.50
4.71	2.	-1
6.28	2.59	-1.5
7.85	1.87	-1.2
9.42	0	-1
10.99	-1.87	-1.23
12.57	-2.59	-1.5
14.14	-2.	-1
15.71	-.866	.5
17.28	-.134	2.23
18.85	0	3

since, if the values are examined carefully, it will be seen that the final point, when $t = 6\pi$, is precisely the same as the initial point when $t = 0$. Thus, for this range of values of t, the resulting shape makes one full cycle to return to its starting point. The above set of points are plotted and connected in Figure 6.4 to show the shape of this particular hypocycloid.

EXERCISE 3

Prepare a version of the program PARAMEQN that can be used exclusively for working with hypocycloids. In particular, provide INPUT statements so that you can supply the two radii without retyping the entire lines 40 and 50 each time.

EXERCISE 4

a. Repeat the process in Example 6.6 to sketch the graph of the hypocycloid with $r_0 = 1$ and $r_1 = 4$ using $n = 16$ points. (Hint: Use the range of values 0 to 8π for t.)

b. Sketch the graph when $r_0 = 1$ and $r_1 = 5$ using $n = 20$ points. (Hint: Use the range 0 to 10π for t.)

c. After examining the three curves, can you deduce any pattern?

FIGURE 6.4

Graph of hypocycloid with radii
$r_0 = 1$ and $r_1 = 3$

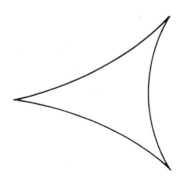

We now consider another application involving parametric equations. The arc length of a curve defined parametrically is given by the integral

$$s = \int_a^b \sqrt{x'^2(t) + y'^2(t)}\, dt$$

As with similar integrals we have seen, this is usually impossible to evaluate in closed form, but can be handled relatively easily using numerical integration. We illustrate this in the following example.

EXAMPLE 6.7   Find the arc length of the ellipse

$$\frac{x^2}{9} + \frac{y^2}{4} = 1$$

At first thought, there seems little need to use parametric equations for this problem. We could simply solve for y as $y = 2\sqrt{1 - (x^2/9)}$ for the upper half of the ellipse, use the usual rectangular formula for arc length to find the length of the upper half of the ellipse, and then double the result. This approach requires using

$$f'(x) = \frac{2(-2x/9)}{2\sqrt{1 - (x^2/9)}} \qquad x \neq 3, -3$$

so that the arc length is given by

$$s = 2\int_{-3}^3 \sqrt{1 + \frac{4x^2/81}{1 - (x^2/9)}}\, dx$$

Unfortunately, the integrand is not defined at either limit of integration, $x = 3$ or $x = -3$, and so the integral cannot be evaluated as it stands. We will consider this problem again in the next chapter when we deal with such integrals.

For now, though, we must seek an alternative approach to solving this problem and so introduce the following parametric representation for the ellipse:

$$x = 3 \cos t$$

$$y = 2 \sin t$$

where t represents the angle from the center of the ellipse to the point (x, y) on the ellipse drawn from the x-axis as shown in Figure 6.5. To see that this parametric representation is valid, we note that

$$x/3 = \cos t \qquad \text{and} \qquad y/2 = \sin t$$

so that

$$x^2/9 + y^2/4 = \cos^2 t + \sin^2 t = 1$$

The corresponding arc length is then given by

FIGURE 6.5

Parameterization of the ellipse

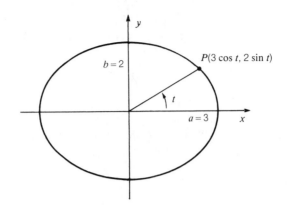

$$s = \int_0^{2\pi} \sqrt{(-3 \sin t)^2 + (2 \cos t)^2}\, dt$$

$$= \int_0^{2\pi} \sqrt{9 \sin^2 t + 4 \cos^2 t}\, dt$$

We note that we used 0 and 2π as the limits of integration, since we want the arc length around the full ellipse and that, further, this integral is well defined for all values of t. Unfortunately, while we have produced a well-defined integral, we are faced with one that cannot be evaluated in closed form. Therefore, we must resort again to a numerical approach to evaluate it. We apply the program SIMP with $n = 1000$ subdivisions (having supplied the function FNY(X) in terms of x rather than t) and so obtain a value of 15.84033 for the arc length.

Before leaving this example, let us consider one other aspect of the problem. Since a circle is a special case of an ellipse, we would expect a formula for the circumference of an ellipse to be similar to the usual formula for a circle, $C = 2\pi r$. In the above example, therefore, we take the answer 15.84033 and divide it by π to obtain 5.04139. Since the lengths of the major and minor half-axes for the ellipse are 3 and 2, respectively, we naturally suspect that the exact arc length should turn out to be 5π and that, in general, $s = (a + b)\pi$.

EXERCISE 5

Obtain a more accurate answer for the above problem by taking a larger value for n in the program SIMP. Does this result support the conjecture?

EXERCISE 6

Repeat the above analysis for the ellipse

$$\frac{x^2}{16} + \frac{y^2}{25} = 1$$

and compare your result to the supposed formula.

SECTION 6.3 PROBLEMS

Use the program PARAMEQN to generate enough points (don't be stingy) to allow you to sketch the following graphs.

1 The hypocycloid with $r_0 = 2$ and $r_1 = 5$.

2 The epicycloid

$$x = (r_0 + r_1) \cos t - r_1 \cos\left(\frac{r_0 + r_1}{r_1}\right) t$$

$$y = (r_0 + r_1) \sin t - r_1 \sin\left(\frac{r_0 + r_1}{r_1}\right) t$$

formed when the rolling circle of radius r_0 rolls around the *outside* of the fixed circle of radius r_1. Use $r_0 = 1$ and $r_1 = 4$.

3 The involute of a circle

$$x = a(\cos t + t \sin t)$$

$$y = a(\sin t - t \cos t)$$

formed by a point P on a string being unwound tautly from about the circle of radius a. Use $a = 1$ and t in $[0, 4\pi]$.

4 The epitrochoid

$$x = r \cos 3Q + R \cos Q$$

$$y = r \sin 3Q + R \sin Q$$

formed when a point S traces out a circular path of radius r and a point P moves as shown in Figure 6.6. Use $r = 5$ and $R = 2$.

5 The path of the endpoint P of a piston rod of length l where it attaches to a circular crank mechanism of radius r is described by the equations

$$x = r^2 - l^2 \sin^2 \phi$$

$$y = l \sin \phi$$

Use $r = 1$ and $l = 3$. (Hint: Find the permissible range for ϕ.) See Figure 6.7.

FIGURE 6.6

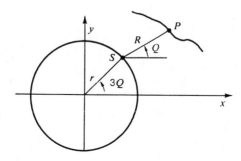

FIGURE 6.7

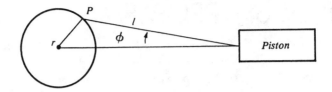

6 Find the arc length of the hypocycloid in Problem 1.

7 Find the arc length of the epicycloid in Problem 2.

8 Find the arc length of the epitrochoid in Problem 4.

SEVEN
INDETERMINATE FORMS, IMPROPER INTEGRALS

7.1
INDETERMINATE FORMS

Expressions of the form $0/0$, ∞/∞, 1^∞, 0^∞, ∞^0, and $\infty - \infty$ often arise when we take the limits of functions. Depending on the function, we find that these expressions can assume any given value whatsoever. For that reason, they are known as *indeterminate forms*.

The primary tool for working with limits involving indeterminate forms is *L'Hôpital's Rule* which says that if

$$\lim_{x \to a} f(x) = \lim_{x \to a} g(x) = 0$$

then

$$\lim_{x \to a} \frac{f(x)}{g(x)} = \lim_{x \to a} \frac{f'(x)}{g'(x)}$$

assuming appropriate conditions on f, g, and g'. A similar result holds if the two limits are both infinite. Limits involving the other indeterminate forms can usually be reduced to or transformed into a $0/0$ or ∞/∞ form.

In this section, we see how the computer can provide some insight into the limits of such functions. For instance, it is fairly easy to show by using the usual procedures for L'Hôpital's Rule that

$$\lim_{x \to 0} (1 + x)^{1/x} = e$$

As a matter of fact, this limit is occasionally used as the definition of e. However, the fact that the number e comes out after some complicated manipulations does not necessarily give someone a feeling that the function does indeed approach e as x approaches 0.

In order to provide the emotional satisfaction to complement the mathematical conviction, we make use of the program LIMFN from Section 2.4. In particular, if we use the function

```
FNY(X) = (1 + X)^(1/X)
```

at $a = 0$ with the sequence of points $h = 1/2^n$, then we obtain the values found in Table 7.1. From these values, clearly the function converges to a

TABLE 7.1

Values for $\lim_{x \to 0}$
$(1 + x)^{1/x}$

N	X - H	F(X - H)	X + H	F(X + H)
1	-.5	4.	.5	2.25
2	-.25	3.160493	.25	2.441406
3	-.125	2.910285	.125	2.565785
4	-.0625	2.808404	.0625	2.637928
5	-.03125	2.762009	.03125	2.676990
6	-.015625	2.739827	.015625	2.697345
7	-.0078125	2.728977	.0078125	2.707739
8	-.0039063	2.723610	.0039063	2.712992
9	-.0019531	2.720941	.0019531	2.715632
10	-.0009766	2.719611	.0009766	2.716955
⋮				
15	-3.0517E-5	2.718333	3.0517E-5	2.718254
⋮				
19	-9.5367E-7	2.718269	9.5367E-7	2.718843

limit of $e = 2.7182818\ldots$ (or something very close to it) before divergence due to round-off errors sets in. See Figure 7.1.

The same approach can be applied to any other limit involving an indeterminate form, as demonstrated in the next example. However, it is important to realize that these results, using the computer in this way, do not constitute rigorous proof of either the value or the existence of the limit. Nonetheless, this approach does provide a strong conviction about what is happening.

FIGURE 7.1

Graph of $y = (1 + x)^{1/x}$
about $x = 0$

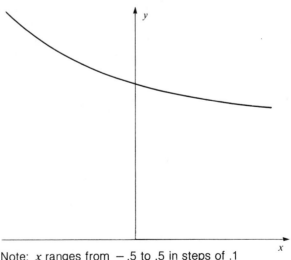

Note: x ranges from $-.5$ to $.5$ in steps of $.1$
y ranges from 0 to 4 in steps of $.4$

EXAMPLE 7.1   Consider

$$\lim_{x \to \pi/2} [\tan x - e^{1/(x-\pi/2)}]$$

We note that this limit produces the indeterminate form $\infty - \infty$. More-over, the limit would be quite difficult to evaluate using the techniques associated with L'Hôpital's Rule. Using the program LIMFN with $a = \pi/2$ and $h = 1/2^n$, we obtain the set of values in Table 7.2. It is evident from these values that the limit does not exist as $x \to \pi/2$.

When we examine Table 7.2, an interesting fact becomes apparent. Each entry listed under F(X − H) is approximately twice the preceding entry, and in fact, each is extremely close to the value for 2^n for $n = 2, 3, \ldots, 10$. That is, $f(\pi/2 - h) \approx 1/h$.

EXERCISE 1   Test the above relationship using other sequences for h, say $h = 1/3^n$ or $h = 1/5^n$ or $h = 1/N^2$.

SECTION 7.1 PROBLEMS

Many indeterminate forms were hidden in previous problem sets, including those for Sections 2.4, 2.5, 2.7, 3.1, 5.1, 5.5, and 5.6. Identify all of them, and interpret the answers you previously obtained in terms of the concept of indeterminate form.

7.2
IMPROPER INTEGRALS

In this section, we will attempt to use the computer to give us some information about the behavior of improper integrals. We begin with a function

TABLE 7.2

Values for $\lim_{x \to \pi/2}$ ·
$[\tan x - e^{1/(x-\pi/2)}]$

N	X − H	F(X − H)	X + H	F(X + H)
0	.570795	.274211	2.570795	-3.360376
1	1.070795	1.695147	2.070795	-9.219549
2	1.320795	3.897980	1.820795	-58.51449
3	1.445795	7.957869	1.695795	-2988.916
4	1.508295	15.97882	1.633295	-8886126.
5	1.539545	31.98822	1.602045	-7.89629E13
6	1.555170	63.98936	1.586420	-6.23515E27
7	1.562983	127.9757		Overflow
8	1.566889	255.9118		
9	1.568842	511.6519		
10	1.569818	1022.611		

$y = f(x)$ that is continuous on $[a, b)$ and such that $\lim_{x \to b} f(x)$ does not exist. The corresponding improper integral is

(1) $\displaystyle\int_a^b f(x)\, dx$

To test such an integral for convergence or divergence, we instead evaluate the proper integral

(2) $\displaystyle\int_a^c f(x)\, dx$

where $c < b$, and then take the limit of the result as c approaches b.

In this approach, the definite integral (2) is expressed in terms of c, and we can then operate on c via the limit process. However, if we want to apply the computer, then we can deal only with numerical values for the quantities involved. This can make things more difficult. Specifically, we can apply a method such as Simpson's Rule to the above integral (2), using a succession of values for c that approach b. In turn, each application of Simpson's Rule will generate a corresponding value for the integral as c approaches b. We expect the limit of this sequence to approximate the value of the original improper integral, if it converges. If this sequence apparently diverges, then we may conclude that the original integral (1) also probably diverges.

We illustrate this approach in the following examples.

EXAMPLE 7.2

In Example 6.7 of Section 6.3, we attempted to find the perimeter of the ellipse $x^2/9 + y^2/4 = 1$, using the usual integration formula for arc length. This led to the definite integral

$$2\int_{-3}^{3} \sqrt{1 + \frac{4x^2/81}{1 - (x^2/9)}}\, dx$$

that is not defined at either endpoint and is therefore an improper integral. We now apply the above procedure using Simpson's Rule to this improper integral. In particular, by making use of the symmetry involved in the ellipse, we need only concern ourselves with the upper right half of the ellipse and multiply the result by 4 to obtain the full perimeter (assuming the integral can be evaluated). Thus, we apply the program SIMP to the function

$$4\sqrt{1 + \frac{4x^2/81}{1 - (x^2/9)}}$$

with $n = 1000$ subdivisions for a succession of values for c that approach 3. The results are shown in Table 7.3. Based on this example, which we know should involve a convergent integral because the arc length is apparently 5π, it is clear that the procedure suggested is very sensitive. As soon as c starts

TABLE 7.3

Computed values for the perimeter of the ellipse $x^2/9 + y^2/4 = 1$

c	s
2.99	15.2111787
2.999	15.6678419
2.9999	15.9584224
2.99999	16.6722047
2.999999	18.9060973
2.9999999	26.002394
2.99999999	48.4082497

to approach 3, the values computed for the integral begin to diverge rapidly, and so we cannot come to a conclusion based on the tabulated results.

We now consider another example of the use of the computer with an improper integral.

EXAMPLE 7.3   Consider the improper integral

$$\int_0^{\pi/2} \sqrt{1 + \tan x}\, dx$$

We note that this integral is not defined at the upper endpoint $x = \pi/2$. Furthermore, it cannot be integrated in closed form using any elementary techniques of integration. Instead, we can attempt to evaluate the definite integral

$$\int_0^c \sqrt{1 + \tan x}\, dx$$

using a succession of values for c that approach $\pi/2 = 1.5707963\ldots$, namely $c = \pi/2 - 1/10^n$ for $n = 1, 2, 3, \ldots$. The corresponding results are shown in Table 7.4. Again, we see that the computed values begin to diverge as soon as c comes anywhere close to the limit of integration.

TABLE 7.4

Computer values for the improper integral $\int_0^c \sqrt{1 + \tan x}\, dx$ as $c \to \pi/2$

c	Simpson
1.4707963	2.24803059
1.5607963	2.69033871
1.5697963	2.82795979
1.5706963	2.88960875
1.5707863	3.00577659
1.5707953	3.35760580
1.5707962	4.31269383
1.57079629	5.59247904
1.57079630	6.00224191

However, since we don't know in advance whether the improper integral converges or diverges, we have no way of concluding anything about the integral.

Up until this section, we have been quite successful in applying the computer to most of the topics in calculus. These two examples, though, demonstrate that things are not always as simple as one might expect. The procedure outlined at the beginning of the section was certainly reasonable — use a numerical integration technique on a succession of definite integrals approaching the desired one. On a small computer, this is a very time-consuming process, but it sounds as if it should work, given sufficient time. Unfortunately, the process does not work particularly well.

To see what went wrong, let's examine the actual formula for Simpson's Rule. The formula states

$$s = \frac{h}{3}[f(a) + 4f(a + h) + 2f(a + 2h) + 4f(a + 3h)$$
$$+ \cdots + 4f(c - h) + f(c)]$$

In particular, we will interpret this in terms of Example 7.2 for the arc length of the ellipse where c approaches 3. The primary problem is that as c nears 3, the term $f(c)$, which occurs explicitly in the formula, introduces an increasingly large contribution to the approximation. (Geometrically, the tangent line to the ellipse near $x = 3$ becomes increasingly vertical.) As a result, the value calculated according to Simpson's Rule cannot be accurate (no matter how small h becomes) once c is near 3 — it involves facing $f(c)$ head on as $c \to 3$. Despite this, we *know* that the perimeter is finite, and so the improper integral must converge. Thus, we are presented with a situation where the computer is not particularly helpful in addressing the calculus.

Before leaving this example, we can make one more point about it. It is possible to attempt to "fix" things in several ways. For instance, we might try to compensate for the increasing size of $f(c)$ as c approaches 3 by selecting a much smaller value for the step h — after all, the actual term in the formula is $f(c)(h/3)$, since each term is multiplied by the factor $h/3$. Again, while theoretically this might work to get us somewhat closer to $x = 3$, the longer time needed to perform the calculations makes this approach not much of a gain. For example, using $n = 1,000,000$ instead of 1000 requires 1000 times the number of calculations, and hence 1000 times the amount of time. Even with the speed of a computer, this will be significant. Worse, the problem will still be there as $c \to 3$. A compromise might be to evaluate the integral first from 0 to 2.99, say, using Simpson's Rule with $n = 1000$ subdivisions, and then to proceed from 2.99 to some point much closer to c, using $n = 1000$ again over this considerably smaller range. This gives a somewhat more accurate result, but the gain may not necessarily be worth the effort. The details of such an approach are left for one of the problems.

Incidentally, there are far more effective and sophisticated methods available for evaluating definite integrals than those we have considered in this book that might produce considerably better results in such a case. See, for example, any textbook on numerical analysis.

SECTION 7.2 PROBLEMS

Apply the program SIMP or one of your modifications of it to determine whether the following improper integrals appear to converge or diverge.

1 $\displaystyle\int_0^e \ln x \, dx$

2 $\displaystyle\int_2^5 \frac{dx}{(x-2)^{1/3}}$

3 $\displaystyle\int_{-1}^1 \frac{dx}{\sqrt{x+1}}$

4 $\displaystyle\int_{-2}^1 \frac{dx}{x+1}$

5 $\displaystyle\int_0^1 \frac{e^x}{\sqrt{x}} \, dx$

6 $\displaystyle\int_0^\pi \frac{\sin x}{\sqrt{x}}$

7 $\displaystyle\int_0^{\pi/4} \frac{\sec x}{x^3} \, dx$

8 $\displaystyle\int_0^1 \frac{e^{-x}}{x^{2/3}} \, dx$

9 $\displaystyle\int_0^2 \frac{\cosh x}{(x-2)^2} \, dx$

10 Attempt to improve the accuracy for the estimate of the perimeter of the ellipse by subdividing the interval $[0, 3]$ into a sequence of successively smaller pieces: $[0, 2.99]$, $[2.99, 2.999]$, $[2.999, 2.9999]$, ... and then applying Simpson's Rule to each. Keep summing the values you obtain in this way so long as they appear to remain reasonable. How close can you get to 5π?

EIGHT
SEQUENCES AND SERIES

8.1
SEQUENCES AND RECURSIONS

We have seen that most of the numerical results obtained from the computer are in the form of a sequence of numbers $\{a_n\} = \{a_0, a_1, a_2, \ldots\}$. In this section, we consider the topic of sequence in its own right, with particular emphasis on the representation of sequences.

When we first introduced the notion of a sequence in Section 2.3, we made a point of trying to determine the general term of the sequence. With an expression for a_n in terms of n, it is possible to write out all terms explicitly and easily. For example, suppose $a_n = 3(2/5)^n$, so that $a_1 = 3(2/5), a_2 = 3(4/25)$, and so on.

Unfortunately, it is not always possible to obtain such an expression for the terms of a sequence. For instance, say we wish to evaluate a definite integral using the Trapezoid Rule. We can apply the rule with $n = 1, 2, 3, 4, \ldots$ subdivisions to obtain a sequence of approximations that will converge to the correct value as $n \to \infty$. Despite this fact, it is hopeless to expect that there will be an explicit formula for the terms of the sequence for any value of n — there simply is no predictable way the terms of the sequence are related to one another or to the number of subdivisions n to generate any form of pattern.

One important situation arises very frequently in practice where a straightforward relationship does exist between the terms of a sequence. Usually, each term in the sequence is expressible in terms of the preceding element. For example, we might have each term being 3 times the preceding one:

$$a_{n+1} = 3a_n$$

If the initial term is known, say $a_0 = 1$, then it is simple to discover that $a_n = 3^n$ for all n. Alternatively, if a_0 were 16, then it would turn out that $a_n = 16(3^n)$. This type of relationship is known as a *recursion relation* or *difference equation*. In many situations, we may be faced with the recursion relation and have little need even to find the general formula for a_n.

It is also possible that a recursion relation exists where each term is specified in terms of two or more of the immediately preceding terms. One of the most famous sequences in mathematics is based on such a recursion relation. It is known as the *Fibonacci Sequence*:

$$1, 1, 2, 3, 5, 8, 13, 21, 34, \ldots$$

This sequence has the property that each term after the second is the sum of the two preceding terms. Thus,

$$a_n = a_{n-1} + a_{n-2}$$

In the 800 years since its discovery, this sequence has been found to have major applications in such apparently diverse areas as economics, psychology, biology, and art, to name just a few. For example, suppose you cut open a spiral seashell and measure the distances from the center to each successive ridge in the spiral. Each such distance (in any direction) is the sum of the two inner distances. Therefore, the successive radii are all multiples of the Fibonacci numbers. (The multiple is just a scale factor.)

The above recursion formula for the Fibonacci Sequence is more commonly written as

$$a_{n+2} = a_n + a_{n+1}$$

with $a_0 = a_1 = 1$ as the initial condition. For readers who are intent on closed form expressions, it is possible to obtain such an answer for this sequence, namely,

$$a_n = \left[\frac{\sqrt{5} + 1}{2\sqrt{5}} \left(\frac{1 + \sqrt{5}}{2} \right)^n \right] + \left[\frac{\sqrt{5} - 1}{2\sqrt{5}} \left(\frac{1 - \sqrt{5}}{2} \right)^n \right]$$

for all $n \geq 0$. The appearance of this expression, though, may be sufficient to give some people doubts about the necessity of always needing a closed form answer to every problem. Nonetheless, it is worth checking that this does give the correct results for $n = 1$ and $n = 2$.

When we consider the complexity of a result such as this one that arises from a very simple recursion relation, it becomes clear that it may be desirable to use an alternative approach. It is particularly natural to use the actual recursion relation itself to generate the terms of such a sequence as needed on the computer, as is done in the following program.

Program RECURS

```
10   REM FIBONACCI TYPE RECURSION RELATIONS
20   INPUT "WHAT IS THE FIRST TERM X0? ";X0
30   INPUT "WHAT IS THE SECOND TERM X1? ";X1
40   INPUT "HOW MANY TERMS DO YOU WANT TO
     SEE? ";N
50   FOR I = 1 TO N
60   LET Y = X0 + X1
70   PRINT I,Y
80   LET X0 = X1
90   LET X1 = Y
100  NEXT I
110  END
```

It is worth noting that lines 80 and 90 are included to update the values being used. That is, once X0 and X1 are used to form Y (= X2), then we no longer need the X0 value, but we do need the X1 value and the Y value to calculate the following term of the sequence. Thus, we set X0 to be the X1 value and reset X1 to be the Y value. This frees Y to be the next term.

If the recursion relationship is other than the Fibonacci equation, then all we need do is modify line 60 of the program to generate the comparable output. For instance, Newton's Method (see Section 3.4) can be interpreted as a recursion relation, and the program NEWTON given there is fairly similar to the program RECURS above.

EXERCISE 1

Modify program RECURS to handle Fibonacci-type sequences based on the sum of the previous three elements.

SECTION 8.1 PROBLEMS

Use a program such as RECURS to calculate the first ten terms of the following recursion relations.

1 $a_{n+1} = \dfrac{a_n}{n + 1}$, $a_0 = 1$

2 $a_{n+2} = -\dfrac{n - 1}{n + 2} a_n$, $a_0 = 1$, $a_1 = 1$

3 $a_{n+3} = \dfrac{2a_n}{(n + 3)(n + 2)}$, $a_0 = 1$, $a_1 = 2$, $a_2 = 3$

4 $a_{n+2} = 5a_{n+1} - 2a_n$, $a_0 = 1$, $a_1 = 4$

5 $a_{n+2} = 5a_{n+1} - 2a_n$, $a_0 = 1$, $a_1 = 3$

6 The limit of the ratios of successive terms in the Fibonacci Sequence, $\lim_{n \to \infty} a_{n+1}/a_n$, is known as the Golden Ratio.

(a) Apply a program such as the one in Exercise 6 in Section 2.3 to the closed form expression for the ratio to obtain a value for the Golden Ratio.

(b) Write a program modifying RECURS that will calculate the ratio of every two successive terms of the Fibonacci Sequence, and use it to obtain a value for the Golden Ratio.

(c) Show via algebraic manipulation of the closed form expression for the Fibonacci Sequence that the limit of the ratio is $(1 + \sqrt{5})/2$. (Hint: Factor out the first term in the numerator and the denominator.)

8.2
SEQUENCES AND SERIES

Suppose we have a sequence of numbers $\{a_n\} = \{a_0, a_1, a_2, \ldots\}$. Such a sequence may converge to a limit L as $n \to \infty$, or it may diverge.

In this section, we consider the problem of summing all the terms of a sequence to produce an *infinite series*

$$\sum_{n=0}^{\infty} a_n$$

Such a sum may itself produce a finite number, and in that case we say that the series is *convergent*. Otherwise, we say it is *divergent*. One useful result is that if the sequence of terms $\{a_n\}$ that is being summed does not converge to 0, then the series must diverge. Thus, if the sequence $\{a_n\}$ diverges or even converges to a value other than 0, the corresponding series (or sum) diverges. However, the simple fact that the sequence of numbers $\{a_n\}$ converges to 0 does not of itself guarantee the convergence of the series Σa_n.

The convergence of a series can often be determined using a variety of tests including the ratio test, the integral test, and the comparison test, among others. At best, use of the computer can provide only a good clue about the convergence or divergence of a series. Since the computer can handle only a finite number of calculations, it can never sum an entire series, but only a fairly large number of terms. Often, this can be helpful in suggesting convergence or divergence, but occasionally the suggestion can be quite misleading.

For example, suppose we consider two well-known series

$$\sum_{n=1}^{\infty} \frac{1}{n^2} \quad \text{and} \quad \sum_{n=1}^{\infty} \frac{1}{n}$$

The first is known to converge (by the integral test); the second is the harmonic series and is known to diverge (by the integral test) despite the fact that the terms approach 0. Table 8.1 displays the values generated from a simple computer program for adding the first k terms of a series. From this table, we see that the values for the convergent series have apparently stabilized at 1.64492102 (actually, from around 60,000 to 100,000, the same value is obtained). At the same time, the values for the harmonic series are

TABLE 8.1

Partial sums of two series

k	$\sum_{n=1}^{k} \frac{1}{n^2}$	$\sum_{n=1}^{k} \frac{1}{n}$
1000	1.64393457	7.48547087
2000	1.64443419	8.17836811
3000	1.6460079	8.58374988
4000	1.64468410	8.87139031
5000	1.64473409	9.09450886
10000	1.64483408	9.78760597
15000	1.64486743	10.1930544
20000	1.64490082	10.4807280
30000	1.64490082	10.8861849
40000	1.64490913	11.1738629
50000	1.64491378	11.3970039
100000	1.64492102	12.0901460

still changing slightly. Of course, this does not constitute proof that the first series is convergent or that the second one is divergent. It is entirely conceivable that the terms of a series will stabilize after the first 20,000,000 terms are summed, say, so that it's just a matter of keeping the program running long enough. Similarly, the first series, if we keep going far enough, might continue to increase, though at a much slower rate. After all, we are dealing with an unending supply of numbers being added on. As a consequence, it should be clear that the computer cannot be used to prove either convergence or divergence of a series.

The following program SERIES can be used to produce the results shown in Table 8.1.

Program SERIES

```
10    REM VALUES OF A SERIES
20    INPUT "HOW MANY TERMS? ";K
30    FOR N = 1 TO K
40    LET A = ...
50    LET S = S + A
60    PRINT N,S
70    NEXT N
80    END
```

At line 40, we need to incorporate an expression for the terms of the series in terms of n. Thus, to handle

$$\sum_{n=1}^{\infty} \frac{1}{n^2}$$

we would type

```
40    LET A = 1/N^2
```

Note, by the way, if a series starts with $n = 0$, then line 30 has to be modified. Similarly, if the terms of a series are not defined when $n = 5$, say, then line 30 has to start with $n = 6$.

It is important to realize that we are not dealing with the series in this program—the series consists of the sum of all the terms for *all* values of n. We are rather looking at only the sum of finite sets of terms: the sum of the first term, the first two terms, the first three terms, and so on. In fact, we are actually generating nothing more than a sequence of numbers:

$$s_1 = \sum_{n=1}^{1} a_n = a_1$$

$$s_2 = \sum_{n=1}^{2} a_n = a_1 + a_2$$

$$s_3 = \sum_{n=1}^{3} a_n = a_1 + a_2 + a_3$$

$$s_4 = \sum_{n=1}^{4} a_n = a_1 + a_2 + a_3 + a_4$$

$$\vdots$$

This sequence is known as the *sequence of partial sums* of the series. When it converges to a limit S, that value is the sum of the series and the series also converges.

In practice, the above program SERIES is overwhelming in the volume of data being generated. The program also runs very slowly because of the time involved in each of the PRINT statements.

EXERCISE 2

Modify the program SERIES to eliminate printing out the results at each line. Instead, compare successive values and have the program end when two successive values S_n and S_{n+1} are within a certain tolerance E of each other, say .0000001. Also, provide a check to see if the values are diverging from one another. (You may want this second check to commence after the first 10 or 20 terms since the first few terms of a sequence may not display the typical pattern for the remainder of the terms.)

In a sense, we have seen two extremes here. One is where all the first 100,000 terms, say, of a sequence of partial sums are printed out. The other is where only an apparent conclusion appears. A more satisfying, intermediate approach might be to have a subsequence, say every 1000th term, printed out as was done in Table 8.1. This is done in the following modification of the program SERIES.

Program SERIES2

```
10    REM SUBSEQUENCE OF THE SEQUENCE OF
      PARTIAL SUMS
20    FOR J = 0 TO 100
30      FOR I = 1 TO 1000
40      LET N = 1000 * J + I
50      LET A = ...
60      LET S = S + A
70      NEXT I
80    PRINT 1000 * J,S
90    NEXT J
100   END
```

EXERCISE 3

Modify program SERIES2 to incorporate the checks on the successive values that you had in Exercise 2.

It is essential to reiterate that the above programs cannot conclusively establish the convergence or divergence of any series. All we need to consider is the harmonic series. It will pass any reasonable numerical test we care to implement on the computer. Thus, for any tolerance E within the range of constants permissible in BASIC, the above program modifications would suggest convergence of the harmonic series since it diverges so slowly. Therefore, the only valid way of determining convergence of a given series is through one of the standard methods such as the comparison, integral, ratio, or alternating series tests.

Once convergence of a series has been established, programs such as the ones above become invaluable. In many applications, the solution to a problem can only be expressed as an infinite series. This is the case with many differential equations where the solution is represented as a power series in the variable x. To utilize the solution in any practical way, however, values for the series must be available. Thus, knowing that such a series converges for particular values of x is only the first step. Thereafter, we must be able to obtain an accurate value for the series. This is where the computer becomes essential, as shown in the following example.

EXAMPLE 8.1

The function given by the power series

$$y = f(x) = \sum_{n=0}^{\infty} \frac{2^n x^{2n+1}}{3 \cdot 5 \cdot \cdots \cdot (2n + 1)}$$

$$= x + \frac{2x^3}{3} + \frac{4x^5}{15} + \frac{8x^7}{105} + \cdots$$

is one solution of the differential equation $y'' - 2xy' - 2y = 0$. What is the value of the solution at $x = 1$ and $x = 1/2$?

The ratio test guarantees that the power series converges for all values of x. The problem is to determine the value of the solution at the given points. If $x = 1$, the series reduces to

$$\sum_{n=0}^{\infty} \frac{2^n}{3 \cdot 5 \cdot \cdots \cdot (2n + 1)} = 1 + \frac{2}{3} + \frac{4}{15} + \frac{8}{105} + \cdots$$

In order to evaluate this series, we would like to apply the program SERIES. However, the program requires a closed form expression for the general term of the series, and that is often difficult to express. A preferable method is to express each term recursively, that is, in terms of the preceding one. In this example, the lead term is $a_0 = 1$, and each successive term can be written

$$a_{n+1} = a_n \left(\frac{2}{2n + 1} \right)$$

We incorporate this into SERIES with the following lines:

```
25   LET A = 1: S = A
40   LET A = A * 2/(2 * N + 1)
```

If we now RUN the program, we obtain the results shown in Table 8.2. From this, we conclude that at $x = 1$, the solution is approximately 2.03007847.

TABLE 8.2

Partial sums for the series in
Example 8.1 with $x = 1$

n	Series
1	1.66666667
2	1.93333333
3	2.00952381
4	2.02645503
5	2.02953343
6	2.03000703
7	2.03007018
8	2.03007761
9	2.03007839
10	2.03007846
11	2.03007847
12	2.03007847

We next consider the result with $x = \frac{1}{2}$. The series reduces to

$$\sum_{n=0}^{\infty} \frac{2^n(1/2)^{2n+1}}{3 \cdot 5 \cdot \cdots \cdot (2n+1)}$$

$$= \sum_{n=0}^{\infty} \frac{2^n(1/2)^{2n+1}}{3 \cdot 5 \cdot \cdots \cdot (2n+1)}$$

$$= \frac{1}{2} + \frac{2(1/2)^3}{3} + \frac{4(1/2)^5}{15} + \frac{8(1/2)^7}{105} + \cdots$$

To apply the modified version of program SERIES, we now use

```
25   LET A = .5: S = A
40   LET A = A * .5/(2 * N + 1)
```

and obtain the results shown in Table 8.3. We therefore conclude that the solution is approximately .592296536.

We note that the convergence was quite rapid for the two particular series considered in this example. Actually, the rate of convergence is very dependent on the magnitude of the value of x. When x is larger than 1 in absolute value, the terms of the series do not approach zero quite so fast,

TABLE 8.3

Partial sums for the series in
Example 8.1 with $x = .5$

n	Series
1	.583333333
2	.591666667
3	.592261905
4	.592294973
5	.592296477
6	.592296534
7	.592296536
8	.592296536

and thus many more terms may be needed until a given degree of accuracy is achieved.

To apply the approach from Example 8.1, we give the following, more sophisticated version of the modified program SERIES.

Program PWRSER

```
10    REM VALUE OF A POWER SERIES
20    INPUT "HOW MANY TERMS? ";K
30    INPUT "WHAT IS THE LOWER LIMIT ON
      SUMMATION? ";K0
40    INPUT "WHAT IS THE INITIAL TERM? ";A
50    LET S = A
60    FOR N = K0 TO K
70    LET A = A * ...
80    LET S = S + A
90    PRINT N,S
100   NEXT N
110   END
```

To use this program, it is necessary to express the recursive relation between the terms at line 70.

| EXERCISE 4 | Modify program PWRSER so that it stops when successive values for the series agree to within a given tolerance E. |

SECTION 8.2 PROBLEMS

Use program SERIES2 or one of your modifications of it to study the following series for possible convergence or divergence. Whenever possible, compare your results from the computer with the definitive results obtained using the standard tests for convergence. (You should have plenty of time to apply the tests while the program RUNs.)

1 $\sum_{n=1}^{\infty} \dfrac{1}{n^2}$

2 $\sum_{n=1}^{\infty} \dfrac{1}{n}$

3 $\sum_{n=1}^{\infty} \dfrac{1}{\ln n}$

4 $\sum_{n=1}^{\infty} \dfrac{n}{(n^2 + 1)^2}$

5 $\sum_{n=1}^{\infty} \dfrac{1}{1 + e^{-n}}$

6 $\sum_{n=0}^{\infty} \dfrac{1}{2 + (\frac{1}{2})^n}$

7 $\sum_{n=1}^{\infty} \dfrac{1}{n^2 + 11n + 30}$

8 $\sum_{n=1}^{\infty} \dfrac{n + \cos n}{n^3 + 1}$

9 $\sum_{n=1}^{\infty} \dfrac{\sqrt{n} \ln n}{n^4 + 1}$

10 $\sum_{n=1}^{\infty} \dfrac{\sin n}{\sqrt{n}}$

11 $\sum_{n=1}^{\infty} \dfrac{\sin n}{n^{3/2}}$

12 $\sum_{n=2}^{\infty} \dfrac{(.9)^n}{\ln n}$

13 $\displaystyle\sum_{n=1}^{\infty} \frac{(-1)^n}{\sqrt[n]{n}}$

14 $\displaystyle\sum_{n=1}^{\infty} \frac{1}{\sqrt[n]{n}}$

15 $\displaystyle\sum_{n=1}^{\infty} \frac{3 + (-1)^{n-1}}{(1.01)^{n-1}}$

For each of the following power series, apply the program PWRSER or your modification of it to approximate its value at the given points.

16 $\displaystyle\sum_{n=1}^{\infty} \frac{(x - 3)^n}{n^3}$ at $x = 4, 3.5$

17 $\displaystyle\sum_{n=0}^{\infty} \frac{2^n x^n}{n!}$ at $x = 1, 2, -2.5$

18 $\displaystyle\sum_{n=1}^{\infty} \frac{n^2}{3^{2n}} (x + 7)^n$ at $x = 2, -2, -10$

19 $\displaystyle\sum_{n=0}^{\infty} \frac{3^n x^{n+1}}{2^{n+1}}$ at $x = -\frac{1}{3}, \frac{1}{2}$

20 $\displaystyle\sum_{n=1}^{\infty} \frac{(-1)^n}{2n(n!)} x^{2n}$ at $x = 1, 5, -10$

8.3
TAYLOR'S THEOREM

One of the most useful mathematical tools developed in calculus is Taylor's Theorem. Among other applications, it provides the techniques needed to evaluate functions to any desired degree of accuracy by approximating each function with an appropriate polynomial. As such, it is the basis for all of the usual trigonometric, exponential, and logarithmic tables, as well as the method used by calculators and computers to evaluate these functions.

Taylor's Theorem states that if a function $f(x)$ is sufficiently differentiable on an interval (a, c) (the first n derivatives exist there) and is continuous on the closed interval $[a, c]$, then

$$f(x) = f(b) + f'(b)(x - b)$$
$$+ \frac{f''(b)(x - b)^2}{2!} + \cdots + \frac{f^{(n)}(b)(x - b)^n}{n!} + R_n$$
$$= \sum_{k=0}^{n} \frac{f^{(k)}(b)(x - b)^k}{k!} + R_n$$

where R_n is the remainder after n terms.

Thus, for example, using $b = 0$, we have

$$e^x = 1 + x + \frac{x^2}{2!} + \frac{x^3}{3!} + \frac{x^4}{4!} + \frac{x^5}{5!} + R_5$$

(Note: See Equation (1) in Section 5.2 where this result was obtained in a totally different manner: $\sin x = x - x^3/3! + x^5/5! - x^7/7! + R_7$ and $\cos x = 1 - x^2/2! + x^4/4! - x^6/6! + R_6$.)

Using $b = 1$, we have

$$\ln x = (x - 1) - \frac{(x - 1)^2}{2} + \frac{(x - 1)^3}{3} - \frac{(x - 1)^4}{4} + R_4$$

In general, the remainder or error term R_n will not be known. In fact, if we did know it, then we would have an exact value for the function rather than an approximation. However, as n increases in size, the accuracy of the approximation will hopefully improve and R_n will approach 0. Thus, in the limit as $n \to \infty$, we obtain

$$f(x) = f(b) + f'(b)(x - b)$$
$$+ \frac{f''(b)(x - b)^2}{2!} + \cdots + \frac{f^{(n)}(b)(x - b)^n}{n!} + \cdots$$
$$= \sum_{k=0}^{\infty} \frac{f^{(k)}(b)(x - b)^k}{k!}$$

This last expression is actually a type of series known as a *power series*. Since it involves a variable x, we have a different series for each value of x. Thus, for some values of x, this series converges and for others it diverges. Moreover, the value of the series when it does converge depends entirely on the value of x — it is indeed a function of x.

To get a feel for the significance of the above formula, we will examine two cases in detail. First, we consider $f(x) = e^x$ on the interval $[0, 2]$ where we use the computer to calculate the values of a succession of the Taylor polynomials (the polynomials of degree n, $n = 0, 1, 2, 3, \ldots$) obtained as the partial sums of the first $n + 1$ terms of the Taylor series. The results are shown in Table 8.4, where the columns represent, respectively,

TABLE 8.4

Taylor series approximations to
$f(x) = e^x$

x	P_2	P_3	P_4	P_5	e^x
0.	1.	1.	1.	1.	1.
.2	1.22	1.221333	1.2214	1.221403	1.221403
.4	1.48	1.490666	1.491733	1.491819	1.491825
.6	1.78	1.816000	1.821400	1.822048	1.822119
.8	2.12	2.205333	2.222400	2.225131	2.225541
1.	2.5	2.666667	2.708334	2.716667	2.718282
1.2	2.92	3.208001	3.294401	3.315137	3.320117
1.4	3.38	3.837334	3.997401	4.042219	4.055199
1.6	3.88	4.562668	4.835735	4.923116	4.953032
1.8	4.42	5.392002	5.829402	5.986866	6.049647
2.	5.	6.333336	7.000002	7.266669	7.389056

FIGURE 8.1

Taylor polynomial approximations to e^x (dashed) with $b = 0$

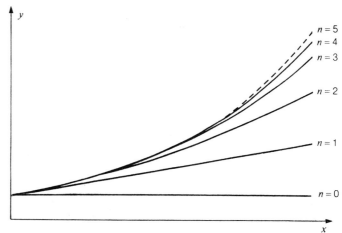

Note: x ranges from 0 to 2 in steps of .2
y ranges from 0 to 10.5664 in steps of 1.05664

$$P_2(x) = 1 + x + \frac{x^2}{2!}$$

$$P_3(x) = 1 + x + \frac{x^2}{2!} + \frac{x^3}{3!}$$

$$P_4(x) = 1 + x + \frac{x^2}{2!} + \frac{x^3}{3!} + \frac{x^4}{4!}$$

$$P_5(x) = 1 + x + \frac{x^2}{2!} + \frac{x^3}{3!} + \frac{x^4}{4!} + \frac{x^5}{5!}$$

and the actual values for e^x (to 6 decimal places). Each of these expressions is then evaluated for each of the values of x shown. The graphs of all of the Taylor polynomials for $n = 0, 1, 2, 3, 4, 5$, as well as the graph of e^x, are superimposed on the same axes in Figure 8.1 for comparison.

In each instance, for a given value of x, we see that each succeeding polynomial presents us with a better approximation to the actual value in the last column. Moreover, the further the x value is from the point $b = 0$, where all the derivatives are evaluated (to obtain the coefficients), the poorer the accuracy for a given polynomial. Thus, to achieve higher accuracy when x is relatively far from b, we must use a higher degree Taylor polynomial.

In Table 8.5 we present a comparable set of values relating to the function $f(x) = \sin x$. The successive columns here represent

$$P_3(x) = x - \frac{x^3}{3!}$$

$$P_5(x) = x - \frac{x^3}{3!} + \frac{x^5}{5!}$$

$$P_7(x) = x - \frac{x^3}{3!} + \frac{x^5}{5!} - \frac{x^7}{7!}$$

and the actual values for sin x. As above, we see that the identical conclusions hold. Graphs of both sin x and the various Taylor polynomial approximations are shown superimposed in Figure 8.2.

TABLE 8.5

Taylor series approximations to $f(x) = \sin x$

x	P_3	P_5	P_7	$\sin x$
0.	0.	0.	0.	0.
.2	.198667	.198669	.198669	.198669
.4	.389333	.389419	.389418	.389418
.6	.564	.564648	.564642	.564642
.8	.714667	.717397	.717356	.717356
1.	.833333	.841667	.841468	.841471
1.2	.912000	.932736	.932025	.932039
1.4	.942667	.987485	.985394	.985449
1.6	.917333	1.004715	.999388	.999574
1.8	.828000	.985464	.973317	.973846
2.	.666667	.933333	.907936	.909297

FIGURE 8.2

Taylor polynomial approximations to sin x (dashed) with $b = 0$

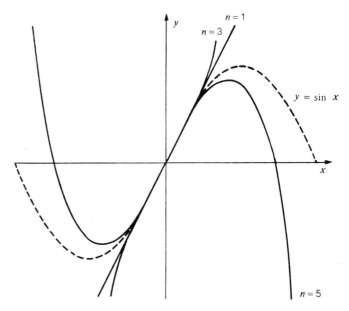

Note: x ranges from -3.14 to 3.14 in steps of .628
y ranges from -1.43 to 1.43 in steps of .286

The following program TAYLOR provides a way of obtaining a table of values for the Taylor approximation to any desired differentiable function.

Program TAYLOR

```
10   REM TAYLOR SERIES APPROXIMATIONS
20   INPUT "WHAT DEGREE POLYNOMIAL DO YOU
     WANT? ";N
30   FOR I = 0 TO N
40   PRINT "WHAT IS A("; I; ")? ";
50   INPUT A(I)
60   NEXT I
70   INPUT "WHAT IS THE RANGE OF VALUES FOR
     X? ";A,C
80   INPUT "AT WHAT POINT B ARE THE
     COEFFICIENTS EVALUATED? ";B
90   LET H = (C - A)/20
100  FOR X = A TO C STEP H
110  LET Y = A(0)
120  FOR I = 1 TO N
130  LET Y = Y + A(I) * (X - B)^I
140  NEXT I
150  PRINT X,Y
160  NEXT X
170  END
```

To use this program, you have to supply the degree of the Taylor polynomial you wish to use and the actual coefficients. Since the INPUT statement can accept only numbers and not operations, you have to calculate the coefficients separately—that is, $1/4!$ would have to be typed in as .041667 and so on. Clearly, any terms that do not appear in the polynomial have coefficients of 0. Further, you have to supply the interval for x and the point b where the terms are evaluated.

EXERCISE 5

Modify the program TAYLOR so that it calculates the factorials recursively. Therefore, the INPUT to the program will consist only of the actual values of the derivatives at the point $x = b$.

As an example, suppose we choose the function $y = \ln x$ evaluated at $b = 2$ using a fourth-degree polynomial. We then have to supply $a(0) = \ln 2 = .693147$, $a(1) = .5$, $a(2) = -.25$, $a(3) = .25$, and $a(4) = -6/16 = -.375$. If the interval selected is $[.5, 2.5]$, then the corresponding output is as shown in part in Table 8.6.

As can be seen, the approximations are very good for values of x near $b = 2$, but become much poorer as x is chosen further away. Figures 8.3 and 8.4 show the successive approximations of the Taylor polynomials to the curve for $y = \ln x$.

TABLE 8.6

Taylor series approximations to
$f(x) = \ln x$ using $b = 2$

x	$P_4(x)$	$\ln x$
.5	-3.361541	-.693147
.7	-1.999641	-.356675
.8	-1.476453	-.223144
.9	-1.041141	-.105360
1.	-.681853	0
1.1	-.387640	.095310
1.2	-.148453	.182322
1.5	.325959	.405465
1.9	.640359	.641854
2.	.693147	.693147
2.1	.740859	.741937
2.5	.888459	.916291

FIGURE 8.3

Taylor polynomial
approximations ($n = 0, 1, 2$)
to ln x (dashed)

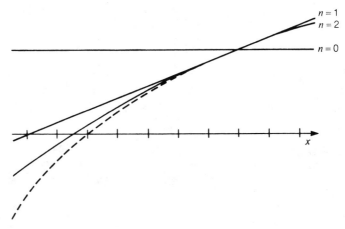

Note: x ranges from .5 to 2.5 in steps of .2
y ranges from $-.991201$ to 1.3103 in steps of .23015

EXERCISE 6

Modify the program TAYLOR so that it also includes the actual values for your choice of function, as was done with Table 8.6. Use either the built-in functions such as SIN, COS, LOG, EXP, and so on, or use a DEF statement.

If the function you want to use is fairly complicated, then evaluating the various derivatives, $f^{(n)}(b)$, can be extremely difficult. One way to avoid this is to make use of program DERIVN2 from Section 3.3 to calculate the values of the derivatives. Recall, though, that the coefficients needed for the Taylor polynomials also include division by the appropriate factorials.

The primary problem with using the methods discussed above is that they are hit or miss, in the sense of giving little feel for the accuracy of the approximation. Thus, we can use the computer to calculate an approximate value for a function $f(x)$ at a point x, but we have no idea if the result is rea-

FIGURE 8.4

Taylor polynomial
approximations ($n = 0, 1, 2, 3, 4$)
to ln x (dashed)

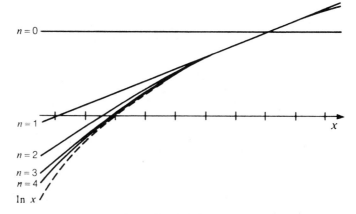

Note: x ranges from .5 to 2.5 in steps of .2
y ranges from $-.991201$ to 1.3103 in steps of .23015

sonably correct or horrendously inexact. We can attack this problem in several ways.

The first method can be applied to the more common functions whose derivatives are fairly simple to find and that form simple recursive patterns. For example, with $f(x) = e^x$, we know that all derivatives are equal to $f^{(n)}(x) = e^x$. Thus, the coefficients in the Taylor series expression are given by

$$a(i) = \frac{e^0}{i!} = \frac{1}{i!}$$

for each value of i. As a result, we can build the Taylor polynomial piece by piece as follows. We start with a value of x, say $x = 1.492$, and let $P = 1 + x$. The next term we need is $x^2/2!$. If we denote the present term by

$$Q = \frac{x}{1}$$

then the next term is

$$Q_1 = \left(\frac{x}{1}\right)\left(\frac{x}{2}\right) = Q\left(\frac{x}{2}\right) = Q\left(\frac{x}{i+1}\right)$$

Therefore, the next polynomial is

$$P_1 = (1 + x) + \frac{x^2}{2!}$$

$$= P + Q_1$$

If the values for P and P_1 are close enough, say,

$$|P - P_1| < .0001$$

then we have achieved the desired accuracy for the approximation. If not, then we reset the present polynomial value

$$P = P_1$$

and reset the present term value

$$Q = Q_1$$

and then construct the next term, $x^3/3!$, as

$$Q_1 = \left(\frac{x^2}{2!}\right)\left(\frac{x}{3}\right)$$

$$= Q\left(\frac{x}{3}\right)$$

$$= Q\left(\frac{x}{i + 1}\right)$$

and so the new polynomial value is

$$P_1 = \left(1 + x + \frac{x^2}{2!}\right) + \frac{x^3}{3!}$$

$$= P + Q_1$$

Again, we compare the successive values P and P_1 to see if they are sufficiently close; if not, we continue the process.

Clearly, this procedure lends itself to a fairly simple computer program. A portion of a program to implement this method for e^x evaluated at $b = 0$ is the following.

```
10    INPUT "VALUE OF X IS ";X
100   LET P = 1
110   LET Q = 1
120   FOR I = 0 TO 20
130   LET Q1 = Q * X/(I + 1)
140   LET P1 = P + Q1
150   IF ABS(P - P1)
          < .0001 OR ABS(Q1) < .0001 THEN 200
160   LET P = P1
170   LET Q = Q1
180   PRINT I,P
190   NEXT I
200   PRINT "THE APPROXIMATION IS ";P1
```

If we apply this program to $f(x) = e^x$ using $x = 1.492$, then we obtain the results shown in Table 8.7. By way of comparison, the correct answer is $e^{1.492} = 4.44597859$.

TABLE 8.7

Taylor series approximations to
$f(x) = e^x$ at $x = 1.492$

i	$P(x)$
0	2.492
1	3.605032
2	4.158580
3	4.365053
4	4.426665
5	4.441986
6	4.445251
7	4.445860
8	4.445961

THE APPROXIMATION IS 4.445976

On the other hand, if the function were $f(x) = e^{2x}$, then each successive derivative would have an extra factor of 2. As a result, the new term Q1 on line 130 would require that extra factor of 2, namely,

```
130   LET Q1 = Q * 2 * X/(I + 1)
```

EXERCISE 7

Modify this program to handle $f(x) = e^{mx}$ for any constant m.

EXERCISE 8

Modify the program to deal with the case where the terms are evaluated at any point $x = b$ instead of at $b = 0$.

EXERCISE 9

Modify the program to handle $f(x) = \sin mx$ for any constant m expanded about $b = 0$. You can account for the changes in sign for successive terms by using $(-1)^I$, or you can set the even-numbered terms to 0 using

```
IF I/2 = INT(I/2) THEN Q1 = 0
```

EXERCISE 10

Modify the program to handle $f(x) = \ln x$ expanded about $x = 1$. Incorporate a check in the program to test whether the subsequent approximations are diverging instead of converging.

As mentioned before, this technique only works nicely when the successive derivatives have a recursive pattern. When they do not, we usually cannot construct the terms as we have done above. One other approach that often works involves having the computer calculate the derivatives of the function as part of the program by using some of the approximation techniques discussed in Chapter 3 and employed, for example, in the program DERIVN2. The problem with this is that the approximation techniques often diverge, so that if we attempt to utilize the values they provide, we may be building in ever more inaccurate results. The final results, therefore, may be grossly inaccurate, and we have no way of recognizing that.

There remains another approach to this problem that involves the use of one of the known formulas for the remainder term R_n. According to the Lagrange form for the remainder after n terms,

$$R_n = f^{(n+1)}(\mathcal{U}) \frac{(x - b)^{n+1}}{(n + 1)!}$$

where the terms are evaluated at $x = b$ and where $\mathcal{U}$ is some unspecified point between b and x. Since we do not know the value for $\mathcal{U}$, we cannot calculate the term $f^{(n+1)}(\mathcal{U})$ exactly. The best we can do is to approximate or replace the value for the $(n + 1)^{st}$ derivative by its maximum value and then to estimate the size of the maximum possible error. For instance, if $f(x) = \sin x$, say, then we know that all derivatives are either $\sin x$ or $\cos x$, so that

$$\left| f^{(n+1)}(\mathcal{U}) \right| \leq 1$$

Therefore,

$$R_n = f^{(n+1)}(\mathcal{U}) \frac{(x - b)^{n+1}}{(n + 1)!} \leq \frac{|x - b|^{n+1}}{(n + 1)!}$$

This expression gives us an upper bound (a greatest possible value) for the error.

In general, if we know that for a given function $f(x)$

$$\left| f^{(n+1)}(x) \right| \leq m$$

then it follows that

$$R_n \leq \frac{m|x - b|^{n+1}}{(n + 1)!}$$

This error expression can be built into the program TAYLOR to provide estimates of the accuracy at each stage. To do so requires supplying the value for m, the maximum value of the $(n + 1)^{st}$ derivative on the indicated interval. We must also construct the factorials up to $n + 1$. This can be accomplished by the program segment

```
30    LET F = 1
40    FOR J = 1 TO N + 1
50    LET F = F * J
60    NEXT J
```

Finally, the values for M * ABS(X − B)/^(N + 1)/F have to be printed.

EXERCISE 11   Modify the program TAYLOR to incorporate this feature.

We illustrate this procedure by applying it to the previous situation with $f(x) = e^x$ at $x = 1.492$. For any n, we know that $f^{(n)}(x) = e^x$, so that

$$\left| f^{(n)}(\mathcal{U}) \right| = \left| e^{\mathcal{U}} \right| \le \left| e^x \right| = e^{1.492} = 4.4459786 = m$$

since e^x is an increasing function on the interval $[0, x]$. Therefore,

$$R_n \le \frac{m|x - b|^{n+1}}{(n + 1)!}$$

$$= \frac{4.445978(1.492)^{n+1}}{(n + 1)!}$$

for any n. We therefore obtain the data in Table 8.8.

In Sections 4.2 and 4.3 on the Trapezoid and Simpson's rules, we were able to take an error estimate and reverse it to determine the value of n needed to assure a desired degree of accuracy. Unfortunately, the comparable procedure cannot be done here in closed form since it is not possible to solve for n explicitly from the error formula

$$E_{\max} = \frac{m|x - b|^{n+1}}{(n + 1)!}$$

due to the presence of the factorial. We could apply several of our earlier programs to calculate the values for the sequence of terms for $E_{\max}$ and so locate the value of n that will produce the desired accuracy. However, our efforts would really be a waste of time since the calculations in the program TAYLOR are performed sequentially for $n = 0, 1, 2, \ldots$, so it would be simple to build in a cut-off when the desired level of accuracy is achieved.

EXERCISE 12

Modify the program TAYLOR to continue until a specified level of accuracy has been reached as measured by the error estimate. Note that this is only possible when you can supply an upper bound on all the derivative terms.

TABLE 8.8

Maximum error estimates for Taylor series approximations

i	$P(x)$	Max Error
0	2.492	6.633399
1	3.605032	4.948516
2	4.158580	2.461062
3	4.365053	.917976
4	4.426665	.2739248
5	4.441986	.068116
6	4.445251	.014518
7	4.445860	.002708
8	4.445961	.000449

SECTION 8.3 PROBLEMS

Modify the program TAYLOR to approximate the value of a function at a single point $x = a$. Then use it to approximate the following quantities. Use the program with several successive values for n to see the convergence. Calculate the error estimate for each approximation.

1 $\sin 46° = \sin(46\pi/180)$ using $b = \pi/4$

2 $\cos 29° = \cos(29\pi/180)$ using $b = \pi/6$

3 $e^{.03}$ using $b = 0$

4 e^{-1} using $b = 0$

5 $\sin 1°$

6 $\cos 3°$

7 $\ln 1.5$ using $b = 1$

8 $\text{Tan}^{-1}(.01)$

Combine the use of programs DERIVN2 and TAYLOR to generate a table of values for the following functions. Use $n = 5$ terms for each.

9 $f(x) = \tan x$ on $[0, \pi/4]$

10 $f(x) = \text{Sin}^{-1}x$ on $[0, 1]$

11 $f(x) = \dfrac{\sin x^2}{x^{3/2}}$ on $[\pi/6, \pi/3]$

Evaluate the following definite integrals using: (a) integration of the Taylor polynomial approximations and (b) Simpson's Rule to compare the two approaches.

12 $\displaystyle\int_0^1 \frac{1 - \cos x}{x^2}\, dx$

13 $\displaystyle\int_0^1 \frac{\sin x}{x}\, dx$

14 $\displaystyle\int_0^{1/2} \frac{\ln(1 + x)}{x}\, dx$

15 $\displaystyle\int_0^3 \frac{\text{Tan}^{-1}x}{x}\, dx$

16 $\displaystyle\int_{-1}^0 \frac{e^{3x-1}}{10^x}\, dx$

NINE
CALCULUS OF TWO VARIABLES

9.1
FUNCTIONS OF TWO VARIABLES

Up until this point, we have dealt exclusively with functions of a single variable in the form $y = f(x)$. We now turn our attention to the case where a quantity z depends on two independent variables, x and y. That is, $z = f(x, y)$.

To apply the computer to such situations, we need two additional features of the BASIC language. The first is an extension of the DEF statement that allows us to handle functions of two variables. The second is known as a *nested loop*.

There are several ways to deal with functions of two variables in BASIC. The most elegant method is to use a two-variable DEF statement in the form

```
DEF FNZ(X,Y) = ...      (any expression in x and y)
```

However, this is available only on some versions of BASIC. To see if it is possible on your version, just type

```
10   DEF FNZ(X,Y) = X + Y
```

and RUN. If you get an error message, then it is not available. Alternatively, you can use a standard one-variable DEF statement in the form

```
DEF FNZ(X) = ...      (any expression in X and Y)
```

This works provided both X and Y have been assigned proper values before the function is called or referred to in a program. We will use the two-variable form in the following discussions. The changes needed to use the alternate form are usually fairly simple.

As an illustration, we might use

```
DEF FNZ(X,Y) = (X^2 + Y^2)/SIN(X + Y)
```

that is defined for all points in the plane except for those where $x + y$ is any multiple of π. Geometrically, the domain excludes all points on the family of parallel lines $y = -x + m\pi$ for any integer m. A more interesting problem is to determine the graph of this function that is some surface in space. We consider this problem later in this section.

The second programming feature we need is known as a *nested loop*. It may take the form

```
FOR A = 1 TO 10
   FOR B = 3 TO 5
   ⋮
   NEXT B
NEXT A
```

The computer reacts as follows. It starts on the outer loop and sets A to be 1. It then proceeds to the next line where it encounters the inner loop. Thus, while $A = 1$, B takes on the values of 3, 4, and 5 successively. Once the instructions with $B = 5$ have been completed and the inner loop is also completed, the program proceeds on to the NEXT A line and is sent back to set $A = 2$. It then proceeds through the entire inner loop again with $B = 3, 4$, and 5 while $A = 2$, and so forth. In essence, this nested loop involves a total of $10 \times 3 = 30$ cycles of operations covering each of the 30 ordered pairs

$$(a, b) = (1, 3), \quad (1, 4), \quad (1, 5)$$

$$(2, 3), \quad (2, 4), \quad (2, 5)$$

$$\vdots \qquad \vdots \qquad \vdots$$

$$(10, 3), \quad (10, 4), \quad (10, 5)$$

We use this structure to generate tables of values for a function of two variables using a program similar to the program TABLE in Section 2.1.

Program TABLE2

```
10   REM TABLE OF VALUES FOR F(X,Y)
20   INPUT "RANGE OF VALUES FOR X ";A,B
30   INPUT "STEP FOR X ";D1
40   INPUT "RANGE OF VALUES FOR Y ";C,D
50   INPUT "STEP FOR Y ";D2
60   FOR X = A TO B STEP D1
70     FOR Y = C TO D STEP D2
80     DEF FNZ(X,Y) = ...
90     PRINT X,Y,FNZ(X,Y)
100    NEXT Y
110    PRINT
120    NEXT X
```

To use this program, you have to define the desired function of x and y at line 80 and use lines 20 through 50 to define the desired *rectangular* domain or base: x from a to b in steps of d_1 and y from c to d in steps of d_2. For example, you might select x from 0 to 3 in steps of .5 and y from 1 to 8 in steps of .25.

It is important to realize that programs involving functions of several variables, or more generally nested loops, tend to take fairly long to RUN. For example, if in the program TABLE2, you asked for 100 steps for both x

and y, then the program would have to process a total of 10,000 repetitions through the loops. As a consequence, the way that the program was set up is not very efficient in the sense that the DEF statement should be located outside the first or outer loop, say at line 15. This would save 10,000 operations from being repeated needlessly. In cases where your computer does not accept FNZ(X, Y), there is an easy way around the problem—instead of using a DEF statement, you simply replace line 80 with an assignment statement in the form

```
80   LET Z = ...
```

using the same expression in terms of x and y. Line 90 also has to be modified so that it reads

```
90   PRINT X,Y,Z
```

In some ways, this might seem to be easier than the DEF statement. For this particular program, it would be. However, if in a certain program we must refer to values of a function at different points within the same context, say $f(x, y)$ and $f(x_1, y_1)$, then this approach would be extremely cumbersome while the DEF method is quite simple.

Before proceeding, it is worth noting that whenever you construct a nested loop, the limits on the variable in the inner loop do not have to be constants, as they were in the program TABLE2. In particular, they can be expressions involving the variable in the outer loop. For example, line 70 of the program can have y ranging between one expression in x and another with an indicated step d_2. Thus, if we want to apply the program TABLE to the function suggested above, $(x^2 + y^2)/\sin(x + y)$, then the use of a rectangular domain is not appropriate. Instead, if we consider the region between $x + y = 0$ and $x + y = \pi$, then we can replace lines 60 and 70 by

```
60   FOR X = .1 TO 3.1 STEP .5
70   FOR Y = -X TO -X + 3.14 STEP .5
```

Note that we "nudged" x away from 0 to avoid the point $x = y = 0$ where the function is not defined.

Once this data is available for a function, it can be used to obtain a feel for the graph of the resulting surface by plotting the points on a three-dimensional Cartesian coordinate system. It can also be used to get a feel for the behavior and special properties of the function. See Figure 9.1.

To generate a good picture of this particular function, however, we must understand what happens in the neighborhood of the point $x = y = 0$. To do this, we have to find the limit, if it exists, of the function as both x and y approach 0. This involves determining the behavior of the function along all possible paths passing through the origin in the xy-plane. To a limited extent, this can be done by hand. For example, along the x-axis with $y = 0$, the function reduces to $x^2/\sin x$. If we rewrite this as $x/(\sin x/x)$ and use the fact that

$$\lim_{x \to 0} \frac{\sin x}{x} = 1$$

FIGURE 9.1

Graph of $z = (x^2 + y^2)/ \sin(x + y)$

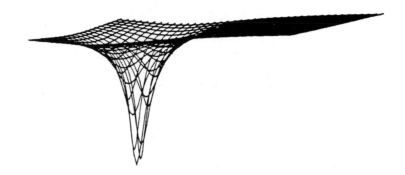

then we see that the limit of the function must be 0 along the x-axis. Similarly, the limit along the y-axis is equal to 0. Further, along any straight line $y = mx$, so long as $m \neq -1$, we obtain

$$f(x, y) = \frac{x^2 + m^2 x^2}{\sin(x + mx)} = \frac{(1 + m^2)x^2}{\sin(1 + m)x}$$

$$= \frac{(1 + m^2)x}{(1 + m)\dfrac{\sin(1 + m)x}{(1 + m)x}}$$

that approaches 0 as x approaches 0. However, to deal with such a situation along other paths becomes a complicated computational procedure that can best be handled using a computer. To accomplish this, we need a program analogous to the program LIMFN of Section 2.4.

Program LIMFN2

```
10   REM LIMITS OF A FUNCTION OF X AND Y
20   INPUT "THE LIMIT POINT A,B IS ";A,B
30   FOR N = 1 TO 20
40   LET X = ...
50   LET Y = ...
60   DEF FNZ(X,Y) = ...
70   PRINT X,Y,FNZ(X,Y)
80   NEXT N
90   END
```

To use this program, you must

1. supply the function of x and y at line 60 (if DEF is not available, use LET Z = ... instead);
2. supply a sequence of x values in terms of n that approach the value a of x at line 40 — say, LET X = A + 1/N^2;
3. supply a functional expression for y in terms of x that passes through the limit point (a, b) at line 50.

We now illustrate the use of the program by applying it to the above function, using the path $y = \sqrt{x}$ that passes through the limit point $(a, b) = (0, 0)$. We use

```
40   LET X = 1/N^2
50   LET Y = SQR(X)
```

in the program LIMFN2 and so obtain the results in Table 9.1. These results also suggest that the limiting value at $(0, 0)$ is 0. This could be seen more clearly if the table were extended or if a sequence of x values had been used that converged to 0 at a faster rate along this path.

On the other hand, suppose we select a different path that also passes through $(0, 0)$, namely $y = -\sin x$. This choice is not made arbitrarily. We know that there is a problem with this function along the line $y = -x$ that has slope $m = -1$. The function $y = -\sin x$ has slope $y' = -\cos x$ that is also -1 at the origin. Therefore, the curve is tangent to the line $y = -x$ as they both pass through the origin. The results are found in Table 9.2. These results clearly indicate that the limiting value obtained along this path is distinctly not equal to 0. In fact, they strongly suggest that the value is not defined along this path. As a consequence, we conclude that the limit for this function does not exist at the origin, and, as a result, the function is not con-

TABLE 9.1

Limit of a function of two variables along $y = x$

n	x	y	$f(x, y)$
1	1	1	2.199500
2	.25	.50	.458454
3	.111111	.333333	.287138
4	.0625	.25	.215998
5	.04	.2	.175009
6	.027777	.166666	.147755
8	.015625	.115	.113220
10	.01	.1	.092004
15	.004444	.066666	.062831
20	.0025	.05	.047760

TABLE 9.2

Limit of a function of two variables along $y = -\sin x$

n	x	y	$f(x, y)$
1	1	-.841471	10.819779
2	.25	-.247404	47.652890
3	.111111	-.110883	107.84471
4	.0625	-.062459	191.91193
5	.04	-.039989	299.94523
6	.0277777	-.027774	431.96438
8	.015625	-.015624	768.14391
10	.01	-.009999	1199.217
15	.004444	-.004444	2700.496
20	.0025	-.0025	4882.604

tinuous there either. Of course, this conclusion can be drawn immediately from the fact that this particular function is not defined along the line $y = -x$, including the point at the origin. Unfortunately, it is not always obvious how to choose a path such as the one just used when we are trying to prove that a limit does not exist.

SECTION 9.1 PROBLEMS

Apply the program LIMFN2 or one of your modifications of it to investigate the possible existence of the limits of the given functions at the indicated points. In each case, use a variety of different paths to approach the limit point.

1 $\dfrac{x^2 - 2}{3 + xy}$ at $(0, 0)$

2 $\dfrac{x^3 - x^2y + xy^2 - y^3}{x^2 + y^2}$ at $(0, 0)$

3 $\dfrac{4x^2y}{x^3 + y^3}$ at $(0, 0)$

4 $\dfrac{xy}{x^2 - y^2}$ at $(1, 1)$

5 $\dfrac{xy}{x^2 - y^2}$ at $(0, 0)$

6 $\ln(2x^2 + y^2)$ at $(0, 0)$

7 $\cos x + \cos y$ at $(0, 0)$

8 $\cos x + 3 \cos(3x + y)$ at $(0, 0)$

9 $-(x^2 + y^{2/3})$ at $(0, 0)$

10 $e^{-x} \sin y$ at $(0, 0)$

11 $e^{-x} \sin y$ at $(0, 2\pi)$

12 $\sin\left(\dfrac{x}{2}\right) \sin y$ at $(0, 0)$

13 $\dfrac{x^2 - 3y^2}{x^2 + 2y}$ at $(0, 0)$

14 $\dfrac{x^3y}{x^6 + y^2}$ at $(0, 0)$

(Hint: Use a cubic path.)

15 $\dfrac{x^2y}{x^4 + y^2}$ at $(0, 0)$

16 $\dfrac{\sin xy}{x^2 + y^2}$ at $(0, 0)$

17 $\dfrac{4x^2 - y^2}{16x^4 - y^4}$ at $(0, 0)$

9.2
PARTIAL DERIVATIVES AND EXTREMUM PROBLEMS

In the present section, we turn our attention to the subject of the partial derivatives of a function $f(x, y)$. These are defined by

$$f_x(x, y) = \lim_{h \to 0} \frac{f(x + h, y) - f(x, y)}{h}$$

and

$$f_y(x, y) = \lim_{k \to 0} \frac{f(x, y + k) - f(x, y)}{k}$$

provided the limits exist. As we have seen, when we take either of the partial derivatives of a given function, we treat the other variable as if it were a constant and apply the usual properties of derivatives. As a consequence, it is reasonable to expect that the numerical methods we used to evaluate the derivative of a function $f(x)$ in Chapter 3 should be applicable here, too. With this in mind, we can use the following program to calculate one of the partial derivatives of $f(x, y)$, based on the definition.

Program PARTIAL

```
10   REM PARTIAL DERIVATIVE OF A FUNCTION
     WITH RESPECT TO X
20   DEF FNZ(X,Y) = ...
30   INPUT "THE POINT WHERE THE DERIVATIVE IS
     TAKEN IS ";A,B
40   FOR N = 1 TO 20
50   LET H = ...
60   LET D = (FNZ(A + H,B) - FNZ(A,B))/H
70   PRINT H,D
80   NEXT N
90   END
```

To use this program, you have to provide the desired function of x and y at line 20 using the DEF statement. If your computer does not accept such a statement, then things become a trifle more complicated. You then need two separate statements, say z_1 for the expression for the function at $(x + h, y)$ and z_2 for the expression at (x, y) at lines 55 and 56. Line 60 then becomes

```
60   LET D = (Z1 - Z2)/H
```

Furthermore, you have to supply the desired point (a, b) where the derivative is to be evaluated and a sequence of values of h that approaches 0 reasonably rapidly, say $h = 1/n^3$. The program produces a table of values from which we can presumably deduce the value for the limit. Of course, the various cautions that applied to the program DERIVF from Section 3.1 apply here as well. If the sequence for h approaches 0 too quickly, then we get no useful information, since the sequence of values calculated for the difference-quotient soon diverges. If h approaches 0 too slowly, then we do not get close enough to a limit to make any clear decision about the value.

EXAMPLE 9.1

Given $f(x, y) = x \operatorname{Tan}^{-1}\sqrt{x/y}$, evaluate f_x at the point $(1.302, 2.338)$.

We use the built-in inverse tangent function in BASIC, ATN, to express the given function in the form

```
DEF FNZ(X,Y) = X * ATN(SQR(X/Y))
```

with the understanding that $y \neq 0$ and the quotient x/y is positive. We now apply the program PARTIAL to this function, using the sequence $h = 1/n^4$, and so obtain the results shown in Table 9.3. Based on these results, we probably would conclude that the most accurate estimate for $f_x(1.302, 2.338)$ would be about .8807. By comparison, after a considerable amount of effort, it turns out that the correct value for the partial derivative is .880756.

EXERCISE 1

Modify the program PARTIAL to compare successive values of d to check if they are within some predetermined tolerance of each other that would indicate the value of the partial derivative. Also, check for when the successive values begin to diverge so that the previous value would be the best approximation. You may want to increase the upper limit on n at line 40.

EXERCISE 2

Modify the program PARTIAL to calculate both partial derivatives at the same time. You may want to use the same sequence h for both f_x and f_y instead of a separate sequence k. In view of the modifications suggested in Exercise 1, you may want to think through whether f_x and f_y should be calculated simultaneously for each value of h or successively (first one completely, then the other).

TABLE 9.3

Approximation of the partial derivative f_x of $x \operatorname{Tan}^{-1}\sqrt{x/y}$

n	h	d
1	1	.964350
2	.0625	.887218
3	.003906	.882050
4	.0016	.881166
5	.0007716	.880924
⋮		
10	.0001	.880763
12	.0000483	.880753
14	.0000260	.880739
15	.0000197	.880717
16	.0000153	.880753
17	.0000120	.880721

Before proceeding, we should note that considerably more effective methods for the numerical calculation of partial derivatives are available, some of which are based on interpolation methods such as we used in Section 3.2. However, we will not go into them here at all.

Probably the most important application of partial derivatives in calculus is the solution of maxima and minima problems for a function of several variables. The usual method involves locating all points where

$$f_x(x, y) = f_y(x, y) = 0$$

and then testing them with the Second Derivative–Discriminant Test. The overwhelming difficulty with applying this method lies in determining the simultaneous solutions of the two (usually non-linear) equations:

$$(1) \qquad \left. \begin{array}{l} f_x(x, y) = 0 \\ f_y(x, y) = 0 \end{array} \right\}$$

For all but the most trivial and artificial cases, this is an extremely difficult task and usually cannot be done in closed form.

The problem actually is similar to the one we faced earlier, when we had to solve for the roots of a single function $f(x) = 0$. To handle that, we employed several different methods, most notably Newton's Method. It turns out that a simple extension of Newton's Method allows us to solve systems of non-linear equations such as (1) above. Before seeing it, let's look at Newton's Method from a slightly different point of view from the one used before. Suppose we have an initial estimate x_0 of the solution to $f(x) = 0$ in one variable. If x_0 is close to the root, then we might seek to find how much must be added to x_0, call it Δx (either positive or negative) to get to the true root. That is, we want to find Δx so that

$$f(x_0 + \Delta x) = 0$$

Now, if x_0 is very close to the root, then Δx is small, and the fact that $f(x)$ is continuous means that

$$f(x_0 + \Delta x) \approx f(x_0) + f'(x_0)\,\Delta x$$

(This follows from the Mean Value Theorem.) We set the right-hand side equal to 0 and solve for Δx to obtain

$$\Delta x = x_1 - x_0 \approx \frac{-f(x_0)}{f'(x_0)}$$

or

$$x_1 = x_0 + \Delta x$$

or

$$x_1 \approx x_0 - \frac{f(x_0)}{f'(x_0)}$$

that is the usual formula for Newton's Method.

We now apply this interpretation to solving the more general system of two equations:

$$(2) \quad \left.\begin{array}{l} f(x, y) = 0 \\ g(x, y) = 0 \end{array}\right\}$$

where f and g are any two functions, not necessarily f_x and f_y. The system of equations (1) above will then just be a special case of this method. Suppose, then, that we start with an initial approximation (x_0, y_0) to the true solution (r, s) and seek to adjust both by appropriate amounts Δx and Δy so that

$$x_0 + \Delta x = r$$

$$y_0 + \Delta y = s$$

Thus,

$$f(r, s) = f(x_0 + \Delta x, y_0 + \Delta y) = 0$$

$$g(r, s) = g(x_0 + \Delta x, y_0 + \Delta y) = 0$$

If Δx and Δy are both small, then it follows that

$$f(x_0 + \Delta x, y_0 + \Delta y) \approx f(x_0, y_0) + f_x(x_0, y_0)\,\Delta x + f_y(x_0, y_0)\,\Delta y$$

$$g(x_0 + \Delta x, y_0 + \Delta y) \approx g(x_0, y_0) + g_x(x_0, y_0)\,\Delta x + g_y(x_0, y_0)\,\Delta y$$

What we want to know are the values for Δx and Δy, the adjustments, that make the left-hand sides both equal to 0. However, if x_0 and y_0 are known, then we also know the values for f, f_x, f_y, g, g_x, and g_y at the point (x_0, y_0), so that the only unknown quantities on the right side are Δx and Δy. This gives us a pair of simultaneous *linear* equations from which we easily find

$$\Delta x = \frac{gf_y - fg_y}{f_x g_y - f_y g_x}$$

$$\Delta y = \frac{fg_x - gf_x}{f_x g_y - f_y g_x} \qquad \text{at } (x_0, y_0)$$

provided that the denominators $f_x g_y - f_y g_x \neq 0$ at this point. Therefore, if (x_0, y_0) is an initial estimate of the solution (r, s), then the next estimate is given by

$$x_1 = x_0 + \Delta x$$

$$y_1 = y_0 + \Delta y$$

When this process is continued repeatedly, we obtain two sequences of values, one for x and the other for y, that converge to the solution extremely rapidly in most instances. Further, this method is particularly well suited for use on a computer, as in the following program.

Program NEWTON2

```
10   REM NEWTON'S METHOD IN TWO VARIABLES
20   INPUT "THE INITIAL ESTIMATE X0, Y0
     IS ";X,Y
30   LET F  = ...
40   LET FX = ...
50   LET FY = ...
60   LET G  = ...
70   LET GX = ...
80   LET GY = ...
90   LET D = FX * GY - FY * GX
100   IF D = 0 THEN 170
110   LET X1 = X + (G * FY - F * GY)/D
120   LET Y1 = Y + (F * GX - G * FX)/D
130   PRINT X1,Y1
140   LET X = X1
150   LET Y = Y1
160   GO TO 30
170   END
```

To use this program, you have to supply the two functions F and G at lines 30 and 60 in terms of x and y, as well as the various partial derivatives at lines 40, 50, 70, and 80. You also have to provide your initial estimate of the solution (usually little more than a guess) via the INPUT statement at line 20.

If you are using the program NEWTON2 to solve a maximum-minimum problem, then be sure that the function F that appears in the program is actually the partial derivative f_x of the original function being optimized. Similarly, the function G in the program is to be the other partial derivative, f_y. Further, the required partial derivatives FX, FY, GX, and GY are just the partial derivatives of f_x and f_y.

EXAMPLE 9.2

Find a critical point for $f(x, y) = x \ln y + x^2 y - 3y^2$.

We have to solve the system of non-linear equations

$$F = f_x = \ln y + 2xy = 0$$

$$G = f_y = x/y + x^2 - 6y = 0$$

In addition, we need the partial derivatives

$$FX = f_{xx} = 2y \qquad FY = f_{xy} = 1/y + 2x$$

$$GX = f_{yx} = 1/y + 2x \qquad GY = f_{yy} = \frac{-x}{y^2} - 6$$

If we now apply the program NEWTON2, starting with an initial guess of (10, 10), say, then we obtain the data in Table 9.4. In fact, the true solution is approximately (.877932707, .452106334), correct to 9 decimal places, and is a saddle point, as seen in Figure 9.2.

TABLE 9.4

Convergence of Newton's Method in two variables

x	y
6.087250	3.828479
3.621006	1.490986
2.008999	.683397
1.099609	.478653
.886145	.452644
.877944	.452106
.877933	.452106
.877933	.452106

FIGURE 9.2

Graph of $z = x \ln y + x^2 y - 3y^2$

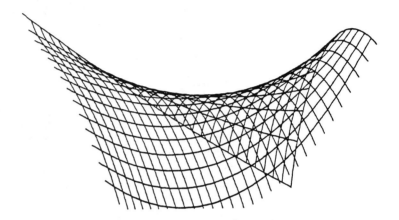

EXERCISE 3

Modify the program NEWTON2 to compare successive values of *both* x and y and to stop when they both get within a predetermined distance of the previous values, say .0001. Also, provide a maximum number of possible iterations, say $n = 20$.

EXERCISE 4

Modify the program NEWTON2 or your modification of it to provide a check to see if either or both sequences start to diverge.

EXERCISE 5

Modify the program NEWTON2 to attempt to salvage the process if d is 0 at line 100. You might attempt to "nudge" the previous values of x and y slightly and try to calculate d again.

SECTION 9.2 PROBLEMS

Apply the program PARTIAL or one of your modifications of it to estimate the values for the following partial derivatives.

1 $f_x(1, 3), f_y(-1, 4)$ for $f(x, y) = x^2 - xy^2 + 4y^5$

2 $f_x(0, 1), f_y(0, 0)$ for $f(x, y) = \dfrac{4\sqrt{x}}{3y^2 + 1}$

3 $f_x(1, 1)$, $f_y(1, 1)$, $f_x(-1, -1)$ for $f(x, y) = e^x \ln xy$

4 $f_x(0, \pi)$, $f_y(0, \pi)$ for $f(x, y) = x \cos(x/y)$

For each of the following problems, locate the extrema for the given function within the domain $[-10, 10] \times [-10, 10]$. First apply the program TABLE2 with a relatively large number of grid points to find the number and approximate location of the extrema. Then apply the program NEWTON2 to calculate the roots of the two partial derivatives correct to four decimal places. Finally, use the derivative test to determine the nature of each of the points you find. Whenever possible, compare these results to the actual values obtained by solving the equations in closed form.

5 $f(x, y) = x^3 + 3xy - y^3$

6 $f(x, y) = x^3 + 4xy - y^3$

7 $f(x, y) = x^4 + y^3 + 32x - 9y$

8 $f(x, y) = x^4 + y^3 + 30x - 5xy$

9 $f(x, y) = (4y + x^2y^2 + 8x)/xy$

10 $f(x, y) = x^3 + y^3 - 6xy + 27$

11 $f(x, y) = x^4 + y^3 - 7xy + 25$

12 $f(x, y) = e^{(y^2 - 3y + x^2 + 4x)}$

13 $f(x, y) = e^{-(y^2 - 3y + x^2 + 4x)}$

14 $f(x, y) = \sin x + \sin y$

15 $f(x, y) = xe^{-x} \sin y$

9.3
MULTIPLE INTEGRALS

We now consider the problem of evaluating the double integral of a function $f(x, y)$ over some region R in the xy-plane, namely,

$$\iint_R f(x, y)\, dA$$

We can treat this as an iterated integral

$$\iint_R f(x, y)\, dy\, dx$$

For simplicity, suppose that R is the rectangular region

$$R = [a, b] \times [c, d]$$

The most logical place to begin our evaluation is with the Riemann Sum, just as we did with the evaluation of single integrals in Section 4.1. For the double integral, the Riemann Sum gives

$$\iint_R f(x, y)\, dx\, dy = \lim_{\substack{\text{all } \Delta x_i \to 0 \\ \Delta y_j \to 0}} \sum_{j=0}^{m-1} \sum_{i=0}^{n-1} f(x_i, y_j)\, \Delta x_i\, \Delta y_j$$

where the base R is subdivided into $m \times n$ subrectangles whose dimensions are $\Delta x_i \times \Delta y_j$ and where (x_i, y_j) is any point in the i,j^{th} rectangle. To make things simpler, suppose there is a uniform spacing $h = \Delta x_i$ horizontally and a uniform spacing $k = \Delta y_j$ vertically. Then,

$$\iint\limits_R f(x, y)\, dy\, dx = \lim_{\substack{h \to 0 \\ k \to 0}} hk \sum_{j=0}^{m-1} \sum_{i=0}^{n-1} f(x_i, y_j)$$

As a final simplification, we assume that the point in each rectangle, (x_i, y_j), is just the lower left corner of that subdivision. As a result, we can write

$$x_i = a + ih \qquad \text{and} \qquad y_j = c + jk$$

for each $i = 0, 1, \ldots, n - 1$ and each $j = 0, 1, \ldots, m - 1$.

This formulation for the Riemann Sum can now be transferred easily into a simple program to approximate the value of a double integral.

Program RMSM2

```
10    REM EVALUATION OF DOUBLE INTEGRALS VIA
      RIEMANN SUM
20    INPUT "THE RANGE OF X VALUES IS ";A,B
30    INPUT "THE RANGE OF Y VALUES IS ";C,D
40    INPUT "THE NUMBER OF SUBDIVISIONS ON X
      IS ";N
50    INPUT "THE NUMBER OF SUBDIVISIONS ON Y
      IS ";M
60    LET H = (B - A)/N
70    LET K = (D - C)/M
80    FOR I = 0 TO N - 1
90    LET X = A + I * H
100     FOR J = 0 TO M - 1
110     LET Y = C + J * K
120     DEF FNZ(X,Y) = ...
130     LET S = S + FNZ(X,Y)
140     NEXT J
150   NEXT I
160   LET S = S * H * K
170   PRINT "THE DOUBLE INTEGRAL IS
      APPROXIMATELY ";S
180   END
```

and then change line 130 accordingly.

In order to use this program, you have to supply the intervals for each variable and the desired number of subdivisions of each, as well as define the desired function at line 120. If your version of BASIC does not accept a function of two variables, you can simply change line 120 to the form

```
120   Z = ...
```

The problem with using this program is that it is grossly inaccurate in most cases, unless n and m are both chosen to be extremely large. For instance, we might want m and n to be at least 100 each and possibly as large as 1000 each. However, bear in mind that a nested loop, such as the one used in this program, involves performing the product of m and n as the total number of operations or cycles. For the numbers we have mentioned, this involves at least 10,000 cycles and possibly as many as 1,000,000 operations to get a reasonably accurate approximation. In turn, this requires a considerable length of time on most small computers, ranging from quite a few minutes up to many hours of continuous operation. This is illustrated in the following example.

EXAMPLE 9.3   Evaluate numerically

$$\int_0^2 \int_1^2 e^{-(x^2+y^2)}\, dy\, dx$$

We note that this type of integral arises very frequently in probability and statistics. Despite this, it cannot be evaluated in closed form by any known method, so we are forced to resort to numerical methods to find its approximate value. To obtain a feel for the effectiveness of the method discussed using the Riemann Sum, we generate a table of values (see Table 9.5) depending on different values for m and n and hence different subdivisions.

Based on these results, we conclude that the value for the definite integral is probably approximately equal to .12. However, if all we can possibly get is two correct decimal places in two hours on a reasonably fast microcomputer, then clearly numerical integration based on the Riemann Sum definition is not really an effective method to use in practice. As a matter of fact, many far more efficient methods have been developed, and we will see one such improvement below.

EXERCISE 6   Modify the program RMSM2 so that the point (x_i, y_j) used in the i, j^{th} rectangle is the center of that rectangle. (This change increases the accuracy considerably for small values of m and n.)

TABLE 9.5

Different subdivisions for a definite integral

m	n	Integral	
5	5	.185784	
5	10	.176411	
10	10	.153751	
20	20	.134973	
25	40	.129314	
30	60	.126953	(5 minutes)
50	100	.123824	(14 minutes)
100	100	.122224	(26 minutes)
200	200	.120748	(2 hours)

EXERCISE 7

Modify the program RMSM2 so that the point (x_i, y_j) used in each rectangle is a randomly chosen point in each rectangle. In particular, use the RND function to select the x and the y values separately.

EXERCISE 8

Modify the program RMSM2 so that it handles the situation where the region R, defined by the limits of integration, is bounded by the two curves $y = g_1(x)$ and $y = g_2(x)$ from $x = a$ to b. In particular, the variable in the FOR-NEXT loops should be x and y with appropriate steps.

In Section 4.2, we found the Trapezoid Rule to be a distinct improvement over the Riemann Sum in evaluating a single integral. The idea there was to approximate the area of a strip under the curve, with the trapezoid formed by a diagonal secant line rather than with the rectangle formed by a horizontal line. See Figure 4.3. When dealing with a double integral, the subdivision of the base into a rectangular array generates a series of rectangular solids of height $f(x_i, y_j)$ for each i and j. The volume of each solid is just $f(x_i, y_j) \Delta x_i \Delta y_j$, and it is these pieces that are added up in the Riemann Sum. Instead of using a horizontal plane as the top of each rectangular solid strip up to the surface, it makes sense to use a slanted plane that is a better fit to the surface above the base rectangle. See Figure 9.3. The resulting shape is prismoidal.

We know that a plane in space is determined by three non-collinear points. Thus, if we pick three points on the surface above the rectangular base subdivision, we will uniquely determine the plane through these points. Without going through the derivation (it appears in Gordon and Greenwell, "The Three-Dimensional Trapezoid Rule," *UMAP Module Unit 425,* COMAP), it turns out that the volume in the i, j^{th} prismoidal approximation is given by

$$V_{ij} = \frac{hk}{2}[f(x_i + h, y_j) + f(x_i, y_j + k)]$$

where we have introduced uniform spacing $h = \Delta x_i$ horizontally and $k = \Delta y_j$ vertically. Therefore, when all the pieces of this approximation are summed, we obtain the formula

$$V \approx \frac{hk}{2}\left[2\sum\sum_{\text{interior}} f(x_i, y_j) + \sum\sum_{\text{boundary}} f(x_i, y_j) \right]$$

with the understanding that the two opposite corners (a, c) and (b, d) must be omitted from the boundary calculation. See Figure 9.4 for a diagram of the subdivision of the base. This result is known as the Three-Dimensional Trapezoid Rule and is a very obvious generalization of the usual trapezoid rule for one variable,

FIGURE 9.3
Approximating plane-based three points

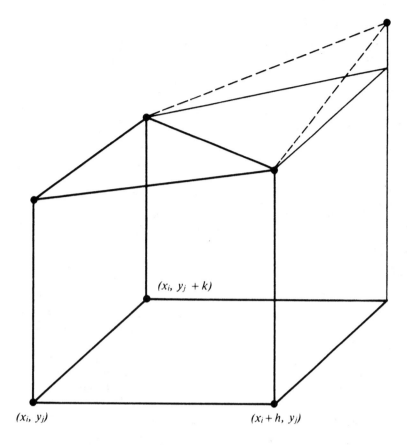

$(x_i, y_j + k)$

(x_i, y_j)

$(x_i + h, y_j)$

FIGURE 9.4
Subdivision of base rectangle R for Three-Dimensional Trapezoid Rule

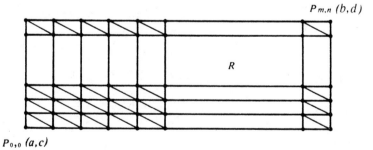

$P_{m,n}\ (b,d)$

R

$P_{0,0}\ (a,c)$

$$\text{Area} \approx \frac{h}{2}\{f(x_0) + f(x_n) + 2[f(x_1) + f(x_2) + \ldots + f(x_{n-1})]\}$$
boundary   interior

Moreover, the results obtained by using the Trapezoid Rule are considerably more accurate than those obtained by using the Riemann Sum for the same number of subdivisions m and n. That is, we get comparable or better accuracy with smaller values of m and n, so the computer time involved is considerably shorter.

TABLE 9.6

The Three-Dimensional
Trapezoid Rule with different
subdivisions

m	n	Integral	
5	5	.128262	
5	10	.127358	
10	10	.122819	
20	20	.120174	
25	40	.119682	
30	60	.119562	(1 minute)
50	100	.119387	(5 minutes)
100	100	.119330	
200	200	.119316	

EXAMPLE 9.4

We illustrate the use of the Three-Dimensional Trapezoid Rule by applying it to the same problem in Example 9.3

$$\int_0^2 \int_1^2 e^{-(x^2+y^2)} \, dy \, dx$$

for the same number of subdivisions. The results are found in Table 9.6. Based on these results, we see that we can get considerably greater accuracy — three digit accuracy of .119 — in a fraction of the time.

A relatively simple program to implement the Three-Dimensional Trapezoid Rule on the computer is as follows.

Program TRAP3D

```
10   REM THREE DIMENSIONAL TRAPEZOID RULE
20   INPUT "THE RANGE OF X VALUES IS ";A,B
30   INPUT "THE RANGE OF Y VALUES IS ";C,D
40   INPUT "THE NUMBER OF SUBDIVISIONS ON X
     IS ";N
50   INPUT "THE NUMBER OF SUBDIVISIONS ON Y
     IS ";M
60   LET H = (B - A)/N
70   LET K = (D - C)/M
80   DEF FNZ(X,Y) = ...
90   REM INTERIOR CALCULATIONS FIRST
100   FOR I = 1 TO N - 1
110   LET X = A + I * H
120     FOR J = 1 TO M - 1
130     LET Y = C + J * K
140     T = T + 2 * FNZ(X,Y)
150    NEXT J
160   NEXT I
170   REM BOUNDARY CALCULATIONS NOW
180   FOR I = 0 TO N
190   LET X = A + I * H
```

```
200    LET T = T + FNZ(X,C) + FNZ(X,D)
210    NEXT I
220    FOR J = 1 TO M - 1
230    LET Y = C + J * K
240    LET T = T + FNZ(A,Y) + FNZ(B,Y)
250    NEXT J
260    REM REMOVE CORNER POINTS
270    LET T = T - FNZ(A,C) - FNZ(B,D)
280    LET T = T * H * K/2
290    PRINT "THE DOUBLE INTEGRAL IS
       APPROXIMATELY ";T
300    END
```

If your version of BASIC does not allow for the function of two variables, then it will be extremely difficult to modify this program without having to retype the expression for the function repeatedly or resorting to a subroutine (which we have not used).

EXERCISE 9   Modify the program TRAP3D so that you can evaluate the Riemann Sum simultaneously and thus compare the accuracy of the two methods.

SECTION 9.3 PROBLEMS

Apply the program RMSM2 or one of your modifications of it to approximate the values for the definite integrals in Problems 1 through 7. Be sure to RUN the program several successive times with different numbers of steps to be able to infer convergence to a value.

Next, apply the program TRAP3D to each of the definite integrals in Problems 1 through 7. Compare the accuracy obtained in each case with the two methods to the number of steps needed and the time required to achieve that accuracy.

1 $\displaystyle\int_0^1 \int_0^1 \frac{x^3 + y^3}{x + y + 1}\, dy\, dx$ **2** $\displaystyle\int_0^1 \int_0^1 \sqrt{x^3 + y^3 + 1}\, dy\, dx$

3 $\displaystyle\int_0^{\pi/2} \int_0^{\pi/2} e^{-\sin(x+y)}\, dy\, dx$ **4** $\displaystyle\int_0^{\pi/2} \int_0^{\pi/2} e^{\sin(x+y)}\, dy\, dx$

5 $\displaystyle\int_{.5}^{2.5} \int_1^2 \frac{xy}{\ln(x + y)}\, dy\, dx$ **6** $\displaystyle\int_{.5}^{2.5} \int_1^2 \frac{\ln(x + y)}{xy}\, dy\, dx$

7 $\displaystyle\int_0^{\pi/2} \int_{\pi/4}^{\pi/2} \frac{\sin x + \sin y}{x + y}\, dy\, dx$

Apply the modification of the program RMSM2 suggested in Exercise 8 to approximate the values for the following definite integrals.

8 $\displaystyle\int_0^1 \int_0^{2x} \frac{x^3 + y^3}{x + y + 1}\, dy\, dx$

9 $\displaystyle\int_0^1 \int_0^{\sqrt{x}} \sqrt{x^3 + y^3 + 1}\, dy\, dx$

10 $\displaystyle\int_0^{\pi/2} \int_0^{\cos x} e^{-\sin(x+y)}\, dy\, dx$

11 $\displaystyle\int_{.5}^{2.5} \int_{e^{-x}}^{e^x} \frac{xy}{\ln(x + y)}\, dy\, dx$

12 $\displaystyle\int_0^{\pi/2} \int_{\mathrm{Tan}^{-1}x}^{x^2/4} \frac{\sin x + \sin y}{x + y}\, dy\, dx$

TEN
COMPUTER GRAPHICS FOR CALCULUS: AN INTRODUCTION

10.1
GRAPH OF A FUNCTION

One of the most exciting areas of computer applications is the field of computer graphics. This is especially true in mathematics where the graphical capabilities of the computer provide us with the opportunity to see many mathematical objects and relationships that are otherwise hidden from view. In many ways, this capability is very much akin to the benefits that microscopes and telescopes have given to physical and biological scientists for centuries—it gives us a view into new worlds.

Based on these comments, you may be wondering why computer graphics have not been a major topic throughout this book. The BASIC statements we have used until now are applicable to virtually all microcomputers and to most time-sharing systems that support BASIC. As a result, almost any student having access to a computer can make full use of all the topics covered. Unfortunately, the additional statements in BASIC that allow for graphics vary dramatically from one machine to another (sometimes even between different models from the same manufacturer) and are often not available at all on time-sharing systems. Consequently, it is literally impossible to have a single book that will do justice to this area without becoming a completely confusing compendium. The other alternative is to have a book dedicated to just one computer model, which leaves everyone else out.

Nevertheless, it would be wrong not to provide at least an introduction to the application of computer graphics to calculus. In this chapter, we include such a discussion. We focus on one particular system of graphics— for the Apple II series of microcomputers. The principles are kept as general as possible, however, and we will afterwards indicate the appropriate changes necessary to implement the same graphics routines on most other common computer models.

The starting point for almost any application of computer graphics in calculus is a program that draws the graph of any desired function $y = f(x)$ on any indicated interval $[a, b]$. Most other graphics applications are just variations and extensions of such a program. Therefore, this introduction to graphics centers on developing this type of program. **199**

Most computer graphics systems basically consider the computer screen as a coordinate grid. The graphics resolution of a computer depends on the number of points available in this grid. For the Apple, the corresponding grid is 280 units wide and 160 units high. These locations are numbered from 0 to 279 horizontally (left to right) and from 0 to 159 vertically (top to bottom). (This latter arrangement with the vertical axis pointing downward in a reversal of the usual Cartesian coordinates is a sad commentary on the lack of impact mathematics has had on the world.) This produces a total of $280 \times 160 = 44{,}800$ possible addressable points that can either be left dark or lit up. Half of the points can be lit in a variety of colors provided you are using a color monitor. See Figure 10.1. A single point whose screen coordinates are H, V (say, 55,111) can be turned on using the statement

```
HPLOT H,V    or    HPLOT 55,111
```

In addition, any two points whose screen coordinates are H1, V1 and H2, V2 can be connected by a straight line (see Figure 10.2) using a variation on the HPLOT statement:

```
HPLOT H1,V1 TO H2,V2
```

A third variation on the HPLOT statement is an instruction to draw a line from the previous point to a new point at H3,V3. This is done with

```
HPLOT TO H3,V3
```

If there are enough points and they are relatively close together, the effect is not a series of straight lines, but rather an apparently smooth curve. The greater the resolution (i.e., the greater the number of addressable points on the screen), the smoother the resulting curve looks. Some of the figures in this book were produced on a special graphics terminal whose resolution is approximately 4000 by 1000 units, with correspondingly finer detail.

FIGURE 10.1

High-resolution graphics page
for Apple II

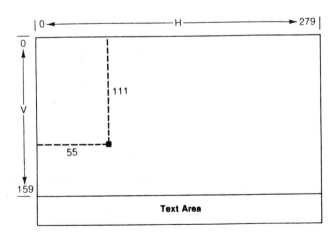

FIGURE 10.2

Connecting points with HPLOT on Apple II

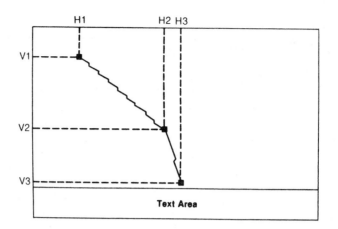

To develop a program that can graph any function, we concentrate on a specific function, namely $f(x) = x^2 + 1$ on the interval $[-1, 2]$. We know that the graph of this function is a portion of a parabola, and, to get the best effect, we want to fit as much of this portion of the curve as possible onto the screen. Thus, we want to restrict our attention to the domain for x between -1 and 2 horizontally and the range for y between 1 and 5 vertically. Essentially, though, we are dealing with two distinct sets of coordinates — the screen coordinates ranging from 0 to 279 horizontally and from 0 to 159 vertically downward — and the user or function coordinates from -1 to 2 horizontally and from 1 to 5 vertically upwards. These two sets of coordinates have to be precisely superimposed on one another to have the desired portion of the graph cover the entire screen area. See Figure 10.3.

We begin the desired program by indicating the screen dimensions:

```
10   LET NH = 279: NV = 159
```

We then need to allow the user to supply any desired function, any desired interval for x from X8 to X9, say, and the size of the horizontal spacing (the total range for x divided by the number of horizontal points NH):

```
50   DEF FNY(X) = X^2 + 1
60   INPUT "WHAT IS THE INTERVAL FOR
     X? ";X8,X9
70   LET DX = (X9 - X8)/NH
```

Since the program we are constructing must work for any function whatsoever, we usually do not know the range for it. Therefore, the program will have to build in a series of instructions that have the computer calculate the largest and smallest values for y on the interval. We let Y8 represent the minimum value for y and Y9 represent the maximum value. Since we are graphing the function at NH (= 279 for the Apple) points, we only have to find the maximum and minimum for the function at the NH points we are

FIGURE 10.3

Fitting a mathematical curve to the graphics page

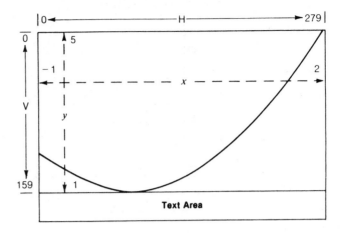

using. This is done as follows. We first set both Y8 and Y9 to the left-most value F(X8) = F(A) and then search across the interval for larger and smaller values of the function.

To do this search, we have to calculate the value of $f(x)$ at NH = 279 uniformly spaced points from $x = a$ to $x = b$ and then compare each to the minima and maxima of Y8 and Y9. Further, when we actually get to the graphing stage in the program, we will need these same NH values of $f(x)$. Therefore, instead of repeating the calculations again at that stage (a needlessly time-consuming job), it makes sense to store them in the computer's memory. The way to accomplish this is to use an *array* Y(I) which consists of a list of NH distinct variables:

```
Y(0), Y(1), Y(2), ... , Y(NH)
```

Each of the NH values calculated for $f(x)$ will be assigned to one of these variables for later use. Moreover, before doing this, it is essential to reserve the necessary space in the computer's memory for all these NH values, and this is done using a DIMension statement near the start of the program. Thus, we insert

```
20   DIM Y(NH)
```

The actual search process is handled by the following set of statements:

```
80    LET Y8 = FNY(X8)
90    LET Y9 = Y8
100   FOR I = 0 TO NH
110   LET X = X8 + I * DX
120   LET Y(I) = FNY(X)
130   IF Y(I) < Y8 THEN Y8 = Y(I)
140   IF Y(I) > Y9 THEN Y9 = Y(I)
150   NEXT I
```

By the time this loop is completed, Y8 has been set to the minimum value for the function and Y9 to the maximum. Also, all major calculations will have taken place (you can expect a delay of about a minute while this happens).

Also, now that we have calculated the range of values for y, we need to subdivide them into the appropriate number of possible vertical locations, NV. This is done by

```
160   LET DY = (Y9 - Y8)/NV
```

We are now ready to begin the actual graphing routine. To do this, we first must instruct the computer to shift into the graphics mode from the text mode. On the Apple, this is accomplished by the HGR command, followed by a statement indicating the color for the points. Thus,

```
170   HGR: HCOLOR = 2
```

The value for the color can be 1 through 8; 1 and 5 are black and will not show up at all on the screen; 4 and 8 are white. The remaining settings give a variety of different hues, depending on the monitor.

Once we are in the graphics mode, we can go on to graph the function. Let's consider first the initial or left-most point on the graph, $x = -1$ and $y = 2$ from $y = x^2 + 1$. The problem is—which point on the screen should be lit up to represent this? Clearly, it should be at the extreme left where H = 0. See Figure 10.4. But what should the vertical screen coordinate V be? Since the vertical range is from 1 to 5, a height of $y = 2$ should fall one-quarter of the way up the screen or three-quarters of the way down from the top. Since the screen coordinates run from 0 at the top to 159 at the bottom, the correct position should be three-quarters of the 159, or about 120, from the top. Specifically, the vertical value V should be $159 - (1/4)159 = 159 - 39.75$. Since the screen coordinates must be integers, it is necessary to round the value to the nearest integer (using the INT statement) to obtain finally $159 - 40 = 119$.

However, if the height can be any possible value y, then the above calculation is somewhat more complicated. We have to compare the value of y to the range from Y8 to Y9, calculate how many steps of length dy that

FIGURE 10.4

Matching screen and mathematical coordinates

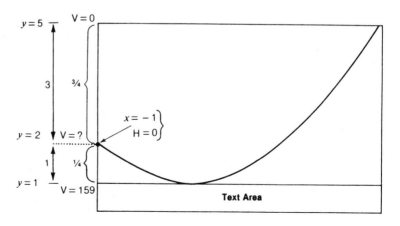

value represents, and subtract the result from 159. In addition, we have to be certain that the result is an integer. Thus, we calculate

```
V = 159 - INT((Y - Y8)/DY + .5)
```

The easiest way to do this is to incorporate these steps into the program itself as a separate function for calculating the vertical screen coordinate for any height y. Further, it is useful to have the same capacity for calculating horizontal positions as well. Thus, we add to the program the following two lines:

```
30   DEF FNH(X) = INT((X - X8)/DX + .5)
40   DEF FNV(Y) = NV - INT((Y - Y8)/DY + .5)
```

Note that since H is measured from 0 at the left to NH = 279 at the right, in line 30 there is no need for any extra term such as the NV = 159 that is required in line 40.

The plotting portion of the program is now quite simple. We first set the initial left-most point and then proceed to connect each point to the succeeding point:

```
200   HPLOT 0, FNV(Y(0))
210   FOR I = 1 TO NH
220   HPLOT TO I, FNV(Y(I))
230   NEXT I
```

That is all there is to it! Of course, some of you may have noticed that lines 180 and 190 were omitted. The only other thing that would be nice is to have the axes drawn in as well. This can present a minor problem. If we were dealing with the same function $Y = X^2 + 1$, but chose instead the interval 100 to 101, then the corresponding range would be 10,001 to 10,202. If we tried to include the x-axis, it would totally distort the shape of the graph because it is so far removed from the portion of the curve we are drawing. The same is true for the y-axis. As a result, it makes sense to include the axes in the graph only if they lie within the portion of the curve being drawn. Thus, the y-axis should appear only if the curve crosses it between X8 and X9. Similarly, the x-axis should be included only if the curve crosses it between Y8 and Y9. We can take care of this with the following two lines:

```
180   IF X8 * X9 <= 0 THEN HPLOT FNH(0),0 TO
      FNH(0),NV
190   IF Y8 * Y9 <= 0 THEN HPLOT 0,FNV(0) TO
      NH,FNV(0)
```

In summary then, the complete graphing program for the Apple II series is the following:

Program FUNCTION GRAPH

```
10   LET NH = 279: NV = 159
20   DIM Y(NH)
30   DEF FNH(X) = INT((X - X8)/DX + .5)
```

```
40    DEF FNV(Y) = NV - INT((Y - Y8)/
      DY + .5)
50    DEF FNY(X) = X^2 + 1
60    INPUT "WHAT IS THE INTERVAL FOR X?
      ";X8,X9
70    LET DX = (X9 - X8)/NH
80    LET Y8 = FNY(X8)
90    LET Y9 = Y8
100   FOR I = 0 TO NH
110   LET X = X8 + I * DX
120   LET Y(I) = FNY(X)
130   IF Y(I) < Y8 THEN Y8 = Y(I)
140   IF Y(I) > Y9 THEN Y9 = Y(I)
150   NEXT I
160   LET DY = (Y9 - Y8)/NV
170   HGR: HCOLOR = 2
180   IF X8 * X9 <= 0 THEN HPLOT FNH(0),0 TO
      FNH(0),NV
190   IF Y8 * Y9 <= 0 THEN HPLOT 0,FNV(0) TO
      NH,FNV(0)
200   HPLOT 0, FNV(Y(0))
210   FOR I = 1 TO NH
220   HPLOT TO I, FNV(Y(I))
230   NEXT I
```

Table 10.1 gives some of the modifications necessary to convert this program for most other popular microcomputer models. In general, the changes involve the screen dimensions NH and NV, the statement to enter High Resolution Graphics Mode, the statement to set a single point, and the statement to connect a given screen point to a succeeding one.

Once you have this program running, it really is the cornerstone of computer graphics for calculus. For example, if you want to examine the behavior of a function, you simply RUN the program repeatedly with different intervals. This is like changing the magnification on a telescope or microscope.

Further, if you want a program that graphs the tangent line to a given curve at a given point, you supply the desired point x_0, calculate the height $f(x_0)$, the slope of the tangent line at that point (just use program DERIVF), find the points where the tangent line leaves the screen region, and connect them using HPLOT, LINE, or a similar statement. Similarly, if you want a program that can illustrate the Riemann Sum for any given function, you supply the number of points (preferably a divisor of NH to make things come out evenly), draw a series of vertical lines from the x-axis to the curve at each of the points, and then draw the series of horizontal lines across the top of each rectangle.

If you want a program to draw curves in polar coordinates, some slight modifications to the above program are necessary. Change the following program lines, and keep all other lines as before.

TABLE 10.1
Table of Graphic Commands for Popular Microcomputers

Model	Dimensions		High Resolution	Point	Line
Apple II	NH = 279	NV = 159	HGR:HCOLOR	HPLOT H,V	HPLOT H,V TO H1,V1
Macintosh	NH = 512	NV = 342		MOVE TO (H,V)	LINE TO (H,V)
IBM PC	NH = 639	NV = 199	SCREEN 2	PSET(H,V)	LINE(H,V) - (H1,V1)
TRS80	NH = 639	NV = 239	SCREEN 0	PSET(H,V)	LINE(H,V) - (H1,V1)
TRS80CC	NH = 255	NV = 191	PMODE 4,1	PSET(H,V)	LINE(H,V) - (H1,V1), PSET
Rainbow	NH = 800	NV = 240	INVOKE DRIVER	MARKER H,V	LINE H,V H1,V1
Atari	NH = 319	NV = 163	GR.8	COLOR:1 PLOT(H,V)	DRAWTO (H1,V1)

```
20    DIM X(200), Y(200)
50    DEF FNR(Q) = ...
52    DEF FNX(Q) = FNR(Q) * COS(Q)
55    DEF FNY(Q) = FNR(Q) * SIN(Q)
60    INPUT "WHAT ARE LIMITS ON Q? ";Q8,Q9
70    LET DQ = (Q9 - Q8)/200
80    LET Y8 = FNY(Q8): Y9 = Y8
90    LET X8 = FNX(Q8): X9 = X8
100   FOR I = 0 TO 200
110   LET Q = Q8 + I * DQ
120   LET X(I) = FNX(Q): Y(I) = FNY(Q)
130   IF Y(I) < Y8 THEN Y8 = Y(I)
135   IF Y(I) > Y9 THEN Y9 = Y(I)
140   IF X(I) < X8 THEN X8 = X(I)
145   IF X(I) > X9 THEN X9 = X(I)
150   NEXT I
160   LET DX = (X9 - X8)/NH: DY = (Y9 - Y8)/NV
200   HPLOT FNH(X(0)), FNV(Y(0))
210   FOR I = 1 TO 200
220   HPLOT TO FNH(X(I)), FNV(Y(I))
230   NEXT I
```

Similarly, a relatively simple modification of this allows you to graph any pair of functions in parametric form. Essentially, all you need to do is remove any reference to R in this last program and replace the two defined functions at lines 52 and 55 with the desired parametric representations.

EXERCISE 1

Modify the above program so that the computer will draw the graph of any curve in parametric form: $x = f(t)$, $y = g(t)$.

Once you have these programs functioning, the nicest thing to do is to just use your imagination to put in the most unlikely functions and see what comes out. Good luck and enjoy!

SECTION 10.1 PROBLEMS

Use the graphing program from this section to produce the graph of each of the following functions on the given interval.

1 $f(x) = x^6 - 7x^5 + 3x^4 + 12x^2 - 15$ on $[-1, 2]$

2 $f(x) = \sin e^x$ on $[-1, 3]$

3 $f(x) = e^{\sin x}$ on $[-3, 6]$

4 $f(x) = \sin(1/x)$ on $[-3, \pi], x \neq 0$

5 $f(x) = \dfrac{\sin x}{x}$ on $[-10, 3\pi]; x \neq 0$

6 $f(x) = (1 - \cos x)/x$ on $[-3, \pi]$; $x \neq 0$

7 $f(x) = x^{\sin x}$ on $[0, 40]$

8 $f(x) = (\sin x)^x$ on $[0, 3.14]$

9 $f(x) = (\sin x)^{\cos x}$ on $[0, 2]$

10 $f(x) = [1 + 1/\sin x]^{\sin x}$ on $[0, 3.14]$

10.2
GRAPHS OF SURFACES IN SPACE

We will now build on the ideas of the previous section on graphing curves to consider the problem of producing computer generated representations of three-dimensional surfaces. This topic is one of the most important applications for computer graphics today and so is the subject of a tremendous amount of time and effort on the part of computer scientists, who are currently developing extremely sophisticated techniques to solve some of the difficult and subtle problems that occur. As a result, we really do little more than scratch the surface of three-dimensional graphical representations.

Suppose we begin with a function $z = f(x, y)$. To simplify matters considerably, we restrict our attention to a case where the domain of the function is a rectangular region R: $[a, b] \times [c, d]$ in the xy-plane. Furthermore, we consider the orientation of the three axes as shown in Figure 10.5 with the x-axis drawn at a 135° angle from both the y- and z-axes to account for the perspective.

With this as a basis, the graph of the function $z = f(x, y)$ will be a surface above (or below) the xy-plane, where the value of z is just the height from the plane to the surface. In particular, we consider only that portion of

FIGURE 10.5

Orientation of three-dimensional coordinate system

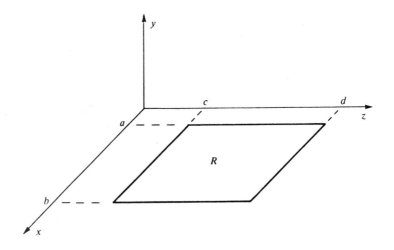

the surface that lies directly above or below the base rectangle R. Our objective is to produce a representation of this portion of the surface using the graphics approach previously discussed.

Suppose we select any particular value of x between a and b, say x_0. With this value fixed, the function $z = f(x, y)$ reduces to a function of the single variable y only. Geometrically, this corresponds to a curve lying on the surface when $x = x_0$. Equivalently, we can think of this curve as being the curve of intersection between the surface and the vertical plane $x = x_0$. In principle, we should be able to instruct the computer to draw the graph of any such curve using the methods of the last section. In practice, as we shall see, several additional factors must be considered.

To produce a reasonably accurate representation of the surface, we repeat the above process for a variety of different values of x, each leading to a different curve lying in the surface. That is, for a set of fixed values of x from a to b, we will generate the corresponding set of curves in the surface, as shown in Figure 10.6.

To graph all these curves representing the surface, we must scale things appropriately so that all the curves fit on the screen. We also have to account for the shift corresponding to the perspective view with the x-axis drawn at the 135° angle.

Suppose we subdivide the x-interval $[a, b]$ into n uniform pieces of width $dx = (b - a)/n$, so that we graph $n + 1$ separate curves. Moreover, suppose that we evaluate each of the curves at $m + 1$ points, $y_0 = c, y_1, y_2, \ldots, y_m = d$ with constant separation $dy = (d - c)/m$. This produces a total of $(n + 1)(m + 1)$ values for the function $f(x, y)$. We keep track of these values in an $n \times m$ array $Z(I, J)$ and use a dimension statement DIM $Z(N, M)$ to reserve the necessary space in memory. Thus,

FIGURE 10.6

Series of contour curves to represent a surface

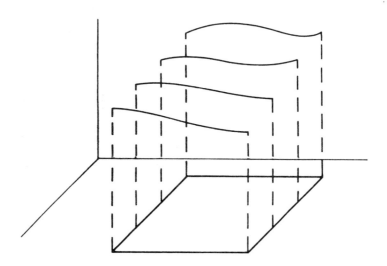

as we did in the curve-sketching program in the previous section, we simultaneously calculate all the functional values and determine the vertical scaling, Y8 and Y9, with the following series of statements:

```
100    DIM Z(N,M)
110    LET Y8 = FNZ(A,C)
or 110    LET Y = C: Y8 - FNZ(A)
120    LET Y9 = M1
130    FOR I = 0 TO N
140    LET X = A + I * DX
150      FOR J = 0 TO M
160      LET Y = C + J * DY
170      LET Z(I,J) = FNZ(X,Y)
or 170      LET Z(I,J) = FNZ(X)
180      IF Z(I,J) < Y8 THEN Y8 = Z(I,J)
190      IF Z(I,J) > Y9 THEN Y9 = Z(I,J)
200      NEXT J
210    NEXT I
```

We now have to shift everything down and to the left to account for perspective. Essentially, we construct a rectangular box from X8 to X9 horizontally and from Y8 to Y9 vertically that just encloses the picture of the portion of the surface we want. The problem is to find the locations for the sides of this box. To see how this is done, we consider how the left-most value, X8, is determined. In the *xyz*-coordinate system we are using, *y* is measured horizontally from *c* to *d*. The left-most point on the screen thus corresponds to some point $S(b,c,z)$ above or below the corner point $P(b,c,0)$ in the base rectangle R. We therefore have to determine how this point can be transformed into X8. It suffices to do the work for the corner point P itself since the identical change will apply to the corresponding corner point on the surface. From Figure 10.7, we see that we must consider

FIGURE 10.7

Shifting a corner point to account for perspective

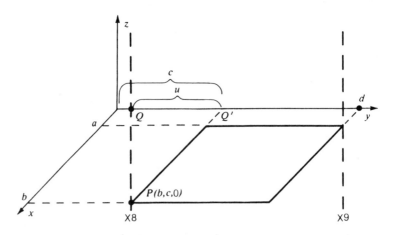

FIGURE 10.8

Close-up view of part of
Figure 10.7

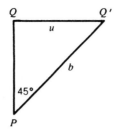

the triangle PQQ' shown in Figure 10.8. From this, it follows that $u = b \sin 45°$, so that $X8 = c - u = c - b \sin 45° = c - b/\sqrt{2}$.

In a similar way, the right-most value is to be X9 and corresponds to the corner point $(a, d, 0)$ of the rectangle R. Therefore, by examining the triangle in Figure 10.9, we conclude that $X9 = d - u = d - a \sin 45° = d - a/\sqrt{2}$.

In totally comparable ways, we shift the maximum and minimum heights also. Thus, starting with the values for Y8 and Y9 that were calculated in the above program segment, it turns out that they have to be modified as follows:

$$Y8 = Y8 - b \cos 45° = Y8 - b/\sqrt{2}$$

$$Y9 = Y9 - a \cos 45° = Y9 - a/\sqrt{2}$$

We include these steps in the graphics program below as lines 220 through 270, where we also set $D1 = (X9 - X8)/NH$ and $D2 = (Y9 - Y8)/NV$.

We now turn to the related question of graphing the curves to represent the surface. If (x, y, z) is any point in space, then it will likewise have to be shifted into screen coordinates the same way that the corner points and the maximum and minimum heights were just done. In particular, it turns out that the transformation is given by

FIGURE 10.9

Shifting other corner points due
to perspective

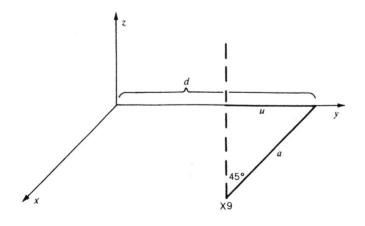

$$\text{Horizontal} = y - x \sin 45° = y - x/\sqrt{2}$$

$$\text{Vertical} = z - x \cos 45° = z - x/\sqrt{2}$$

as seen in Figure 10.10.

Finally, it is necessary to convert the above coordinates into screen coordinates according to the graphics system of the particular microcomputer being used. Again, we let NH and NV represent the number of horizontal and vertical points on the screen and introduce the two functions FNH(X) and FNV(Y). With these, the graphing routine becomes fairly straightforward:

```
300   HGR: HOME: REM GRAPHICS SCREEN
310   FOR I = 0 TO N
320   LET X = A + I * DX
330   LET H = FNH(C - S * X)
340   LET V = FNV(Z(I,0) - S * X)
350   HPLOT H,V: REM SET THE INITIAL POINT
360    FOR J = 0 TO M
370    LET Y = C + J * DY
380    LET H = FNH(Y - S * X)
390    LET V = FNV(Z(I,J) - S * X)
400    HPLOT TO H,V
410    NEXT J
420   NEXT I
```

We now collect all the pieces of this program and put them together. Thus, the complete program becomes:

Program SURFACE

```
10   LET NH = 279: NV = 159
20   DEF FNH(X) = INT((X - X8)/D1 + .5)
30   DEF FNV(Y) = NV - INT((Y - Y8)/D2 + .5)
```

FIGURE 10.10

Shifting an arbitrary point $P(x, y, z)$ due to perspective

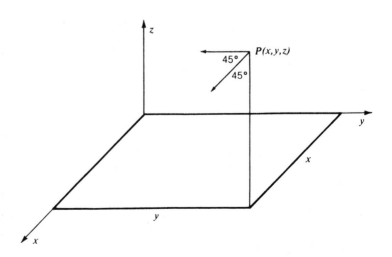

```
 40    DEF FNZ(X,Y) = ...
 50    INPUT "WHAT IS THE RANGE OF VALUES FOR
       X? ";A,B
 60    INPUT "WHAT IS THE RANGE OF VALUES FOR
       Y? ";C,D
 70    INPUT "HOW MANY CURVES DO YOU WANT? ";N
 80    INPUT "HOW MANY Y POINTS? ";M
 90    LET DX = (B - A)/N: DY = (D - C)/M
100    DIM Z(N,M)
110    LET Y8 = FNZ(A,C)
120    LET Y9 = Y8
or 110 LET Y = C: Y8 = FNZ(A)
130    FOR I = 0 TO N
140    LET X = A + I * DX
150     FOR J = 0 TO M
160      LET Y = C + J * DY
170      LET Z(I,J) = FNZ(X,Y)
or 170   LET Z(I,J) = FNZ(X)
180      IF Z(I,J) < Y8 THEN Y8 = Z(I,J)
190      IF Z(I,J) > Y9 THEN Y9 = Z(I,J)
200     NEXT J
210    NEXT I
220    LET S = 1/SQR(2)
230    LET X8 = C - S * B
240    LET X9 = D - S * A
250    LET Y8 = Y8 - S * B
260    LET Y9 = Y9 - S * A
270    LET D1 = (X9 - X8)/NH: D2 = (Y9 - Y8)/NV
300    HGR: HOME: REM GRAPHICS SCREEN
310    FOR I = 0 TO N
320    LET X = A + I * DX
330    LET H = FNH(C - S * X)
340    LET V = FNV(Z(I,0) - S * X)
350    HPLOT H,V: REM SET THE INITIAL POINT
360     FOR J = 0 TO M
370      LET Y = C + J * DY
380      LET H = FNH(Y - S * X)
390      LET V = FNV(Z(I,J) - S * X)
400      HPLOT TO H,V
410     NEXT J
420    NEXT I
```

In Figure 10.11, we show the results of applying the program SURFACE to the function $z = \sin(x^2 + y^2)/(x^2 + y^2)$ with $n = 25$ and $m = 25$.

Before using this program, several cautions are in order. First of all, we are essentially dealing with n curves instead of the single curve handled by the program in the last section. As a result, this program takes quite a lot of time to RUN. With that in mind, the values for n and m should be chosen

FIGURE 10.11

Computer representation of
the surface z =
sin(x² + y²)/(x² + y²)

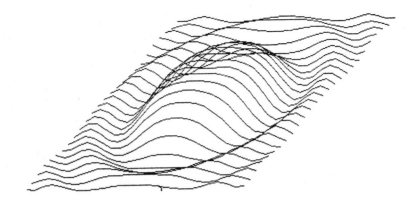

carefully. For instance, the number of curves n might typically be on the order of 20 to 30. Further, the number of points m for y should not be as high as the 279 used in the last section. A value of 20 to 30 is more appropriate. Even with these rather minimal values, the program would involve evaluating the function at a minimum of $20 \times 20 = 400$ points and so would take somewhat longer to RUN than the curve-graphing program, which requires only 279 evaluations.

Still another possible problem involves the choice of interval for either x or y. If you pick an inappropriate range of values for one or both of the variables, the resulting portion of the surface may not have any interesting details—it may appear to be simply a flat plane.

A good strategy in using this program is to RUN it with relatively small values for n and m (say 5 or 6) until you see that you have chosen a good region of the surface to examine by your choices of a, b, c, and d. At that point, you can magnify the detail by increasing the sizes of n and m somewhat to improve the resolution of the resulting image. If the result is still good, you can then go for a higher resolution with large values for n and m to get a very detailed picture. However, this final version can take several hours to produce, so be sure you can monopolize a computer for that long before you RUN the program. Also, be aware that increasing the number of curves used does not automatically give a better image. You will find that with too many curves, detail quickly becomes covered over.

EXERCISE 2

Modify the program SURFACE to include a routine for graphing the three axes. In particular, consider lines 180 through 190 of the program in the last section. Be careful to determine the point where the x-axis leaves the lower part of the screen.

The approach suggested above can be improved on to get better images of the surface. Possibly the simplest improvement is to consider graph-

FIGURE 10.12

Another representation of the surface $z = \sin(x^2 + y^2)/(x^2 + y^2)$

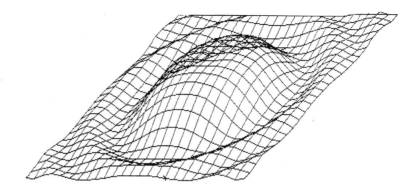

ing a series of curves lying in the surface in both the y-direction (as we did above) and the x-direction. For the latter, we may also consider selecting a series of values for y and considering the resulting functions of x only. This can be implemented with relatively minor changes to the above program. In particular, we need to INPUT the number of x-curves and the number of points on each. Alternatively, we can just use the same number of curves m and points n on each. It is then necessary to introduce a second series of graphics statements paralleling those in lines 300 through 420, where the roles of i and j are interchanged. The result is a graph whose surface looks like it is made of wire-mesh. Such a graph is shown in Figure 10.12 for the function $z = \sin(x^2 + y^2)/(x^2 + y^2)$ with $m = n = 25$.

___

EXERCISE 3

Modify the program SURFACE so that a wire mesh grid effect is produced.

___

EXERCISE 4

Modify the program SURFACE to permit different perspectives on the view of the surface by using angles other than 45°. In particular, use an INPUT statement to supply the desired angle α and modify lines 220 through 260 to include $\sin \alpha$ and $\cos \alpha$.

As mentioned at the beginning of this section, we have only touched on the notions of graphing surfaces. One of the disadvantages of the program SURFACE is that no thought has been paid to "hidden lines." These are the portions of the curves that lie "behind" other curves. When they are drawn, they tend to distort the appearance of the surface. Therefore, it is definitely desirable to draw the representation in such a way that no such hidden lines are shown. A variety of techniques to accomplish this have been developed. The interested reader is referred to any book on computer graphics for details on such methods.

SECTION 10.2 PROBLEMS

Apply the surface-graphing program to each of the following functions. Determine an appropriate region R by experimenting to obtain interesting surface shapes. With each, try different numbers of curves and points.

1 $z = 2x^2 - 2y^2 + 4$

2 $z = \ln(x^2 + y^2 + .005)$

3 $z = \sin xy$

4 $z = e^{-x} \sin y$

5 $z = \cos x + \cos y$

6 $z = \cos x + 3 \cos(3x + y)$

7 $z = \dfrac{\cos(2x^2 + y^2)}{1 + 2x^2 + y^2}$

8 $z = x^{(\sin x + \cos y)}, x \geq 0$

APPENDIX
INTERPOLATING
POLYNOMIALS

In Section 3.2, we found that it was necessary to use a method based on interpolation to obtain an accurate numerical value for the derivative of a function. In this appendix, we consider the idea of interpolation in somewhat more detail and, in a broader sense, consider the notion of approximating one function by another. This type of problem is one that is central to almost all areas where mathematics is applied.

The approach to be developed applies to two different situations. In the first, a function $f(x)$ is known, but is too complicated for our purposes, and we seek to approximate it with a simpler function. Specifically, we select a set of values from the function (geometrically, a set of points on its graph) and produce an approximating or interpolating polynomial that assumes these values (passes through these points). This is illustrated in Figure A.1, where $f(x)$ is the indicated function, P_0, P_1, and P_2 are three points on its graph, and $P(x)$ is the interpolating polynomial (a quadratic) passing through the three points. Thus, at any of the three so-called interpolating points x_0, x_1, and x_2, the function and the polynomial agree exactly, and

$$f(x_0) = P(x_0) \qquad f(x_1) = P(x_1) \qquad f(x_2) = P(x_2)$$

For any value of x other than x_0, x_1, and x_2, we take the value of $P(x)$ as the approximation to $f(x)$.

The second situation that can also be handled by the method to be developed is where there is no explicit function $f(x)$, but rather just a set of measurements of values $P_0(x_0, y_0), P_1(x_1, y_1)$, and $P_2(x_2, y_2)$, and we seek a function $P(x)$ that passes through these points. Once this is done, for any value of the independent variable x other than the value at one of the interpolating points $x_0, x_1, \ldots$, we can use the value of $P(x)$ as the value for the actual function—that is, we can essentially pretend that the interpolating polynomial found is the actual function.

Suppose now that we start with two interpolating points x_0 and x_1 and the corresponding measurements (x_0, y_0) and (x_1, y_1). Obviously, these two points determine a line whose equation is obtained using the point-slope formula as:

$$y - y_0 = \frac{y_1 - y_0}{x_1 - x_0}(x - x_0)$$

FIGURE A.1

Quadratic interpolating polynomial versus a function being
approximated

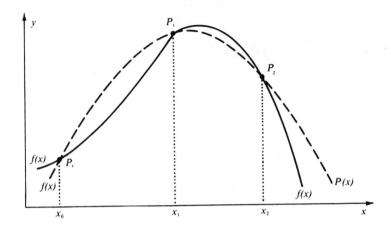

We solve this for y as follows:

$$y = y_0 + \frac{y_1 - y_0}{x_1 - x_0}(x - x_0)$$

$$= y_0\left(1 - \frac{x - x_0}{x_1 - x_0}\right) + y_1\left(\frac{x - x_0}{x_1 - x_0}\right)$$

$$= y_0\left[\frac{(x_1 - x_0) - (x - x_0)}{x_1 - x_0}\right] + y_1\left(\frac{x - x_0}{x_1 - x_0}\right)$$

$$= y_0\left(\frac{x_1 - x}{x_1 - x_0}\right) + y_1\left(\frac{x - x_0}{x_1 - x_0}\right)$$

$$= y_0\left(\frac{x - x_1}{x_0 - x_1}\right) + y_1\left(\frac{x - x_0}{x_1 - x_0}\right)$$

This final form gives the interpolating polynomial of degree 1 passing
through the two given points. It is known as the *Lagrange interpolating
formula* of degree 1,

$$P_1(x) = y_0\left(\frac{x - x_1}{x_0 - x_1}\right) + y_1\left(\frac{x - x_0}{x_1 - x_0}\right)$$

This is shown in Figure A.2. We notice that if x is one of the values x_0 or x_1,
say $x = x_0$, then one of the two terms is zero and the other produces simply

$$P_1(x_0) = y_0 = f(x_0)$$

If x is any value other than x_0 or x_1, then $P_1(x)$ yields an approximating
value.

Just as two points uniquely determine a line, three points (that do not
lie on a line) uniquely determine a parabola. Thus, if we choose three *non-
collinear points* (x_0, y_0), (x_1, y_1), and (x_2, y_2), then we determine precisely
one quadratic function. We select the following form for it:

FIGURE A.2

Linear interpolation

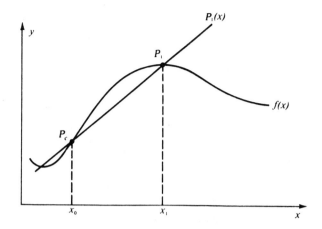

$$P_2(x) = c_0(x - x_1)(x - x_2) + c_1(x - x_0)(x - x_2)$$
$$+ c_2(x - x_0)(x - x_1)$$

where c_0, c_1, and c_2 are three coefficients that must be determined. We note that when $x = x_0$, the polynomial $P_2(x)$ must give the correct value y_0 since it passes through the point (x_0, y_0). However, in the above expression, two of the three terms contain the factor $(x - x_0)$, so when $x = x_0$, these terms will be zero. Hence,

$$P_2(x_0) = y_0 = c_0(x_0 - x_1)(x_0 - x_2)$$

so that

$$c_0 = \frac{y_0}{(x_0 - x_1)(x_0 - x_2)}$$

In an analogous way, we use the fact that $P_2(x_1) = y_1$ to determine

$$c_1 = \frac{y_1}{(x_1 - x_0)(x_1 - x_2)}$$

and the fact that $P_2(x_2) = y_2$ to find

$$c_2 = \frac{y_2}{(x_2 - x_0)(x_2 - x_1)}$$

Thus, the Lagrange interpolating polynomial of degree 2 becomes

$$P_2(x) = y_0\frac{(x - x_1)(x - x_2)}{(x_0 - x_1)(x_0 - x_2)} + y_1\frac{(x - x_0)(x - x_2)}{(x_1 - x_0)(x_1 - x_2)}$$
$$+ y_2\frac{(x - x_0)(x - x_1)}{(x_2 - x_0)(x_2 - x_1)}$$

We now illustrate the above formulas before developing the general case of the Lagrange interpolating polynomial of degree n.

EXAMPLE A.1

Find the Lagrange interpolating polynomial of degree 1 that approximates $f(x) = \sin x$ at the two interpolating points $x_0 = 0$ and $x_1 = \pi/6$.

We immediately have $y_0 = f(x_0) = \sin 0 = 0$ and $y_1 = f(x_1) = \sin(\pi/6) = 1/2$. Therefore,

$$P_1(x) = 0\left(\frac{x - x_1}{x_0 - x_1}\right) + \frac{1}{2}\left(\frac{x - x_0}{x_1 - x_0}\right)$$

$$= \frac{1}{2}\left[\frac{x - 0}{(\pi/6) - 0}\right]$$

$$= \frac{3}{\pi}x$$

Figure A.3 pictures the situation, and Table A.1 presents some typical numerical values.

EXAMPLE A.2

Find the Lagrange interpolating polynomial of degree 1 that passes through the points $(0, 2)$ and $(1, 5)$.

We are given $x_0 = 0, x_1 = 1, y_0 = 2$, and $y_1 = 5$. Therefore,

$$P_1(x) = 2\left(\frac{x - 1}{0 - 1}\right) + 5\left(\frac{x - 0}{1 - 0}\right)$$

$$= -2(x - 1) + 5x$$

EXAMPLE A.3

Find an approximation to the function $f(x) = \log_{10} x$ using the values $x = 1, 1.1,$ and 1.2.

FIGURE A.3

Lagrange interpolating polynomial of degree 1 versus
$f(x) = \sin x$

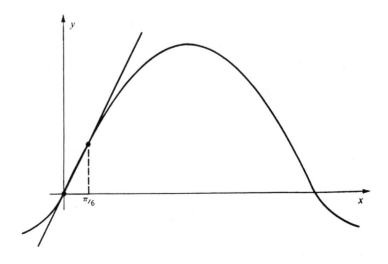

TABLE A.1

Approximations to sin x using interpolation

x	sin x	$P_1(x)$
.0	.0000	.0000
.1	.09998	.0955
.2	.199	.191
.3	.295	.286
$\pi/20$	.156	.150
$\pi/10$	.544	.300

The solution to this problem is obtained by using the Lagrange interpolating polynomial of degree 2 with $x_0 = 1$, $x_1 = 1.1$, and $x_2 = 1.2$. Consequently, we have

$$y_0 = \log_{10}(x_0) = \log_{10}(1) = 0$$

$$y_1 = \log_{10}(1.1) = .0414$$

$$y_2 = \log_{10}(1.2) = .0792$$

Therefore,

$$P_2(x) = 0\frac{(x - 1.1)(x - 1.2)}{(1 - 1.1)(1 - 1.2)} + .0414\frac{(x - 1)(x - 1.2)}{(1.1 - 1)(1.1 - 1.2)}$$
$$+ 0.0792\frac{(x - 1)(x - 1.1)}{(1.2 - 1)(1.2 - 1.1)}$$
$$= -4.14(x - 1)(x - 1.2) + 3.96(x - 1)(x - 1.1)$$

To illustrate the effectiveness of this approximation, Table A.2 lists some sample values.

We now consider the more general situation where we have a set of $n + 1$ interpolating points $x_0, x_1, \ldots, x_n$ and the $n + 1$ corresponding points or measurements $(x_0, y_0), (x_1, y_1), \ldots, (x_n, y_n)$. Just as two points determine a line and three points a parabola, $n + 1$ points will most likely determine a polynomial of degree n. Of course, it is possible that these points will all lie on a curve of lower degree. For instance, it is conceivable that they might all lie on a single line or a parabola or so forth, and so a lower-degree polynomial is all that would be determined.

TABLE A.2

Approximations to $\log_{10} x$ using interpolation

x	$\log_{10} x$	$P_2(x)$
1.05	.02118	.02115
1.15	.06070	.06075
1.25	.09691	.09675
1.30	.11394	.11340
1.35	.13033	.12915
1.40	.14613	.14400

To develop the corresponding Lagrange interpolation formula for the case of $n + 1$ points, we first introduce the following notation with $n = 2$. Let

$$L_0(x) = \frac{(x - x_1)(x - x_2)}{(x_0 - x_1)(x_0 - x_2)}$$

$$L_1(x) = \frac{(x - x_0)(x - x_2)}{(x_1 - x_0)(x_1 - x_2)}$$

$$L_2(x) = \frac{(x - x_0)(x - x_1)}{(x_2 - x_0)(x_2 - x_1)}$$

In each of these, the factors corresponding to the number of the term are left out. For example, $(x - x_1)$ and $(x_1 - x_1)$ are omitted in $L_1(x)$, and so on. The resulting formula for $P_2(x)$ can then be written

$$P_2(x) = y_0 L_0(x) + y_1 L_1(x) + y_2 L_2(x)$$

To extend this to the case of $n + 1$ points, we introduce the following expressions. For any $k = 0, 1, \ldots, n$, let

$$L_k(x) = \frac{(x - x_0)(x - x_1) \cdots (x - x_{k-1})(x - x_{k+1}) \cdots (x - x_n)}{(x_k - x_0)(x_k - x_1) \cdots (x_k - x_{k-1})(x_k - x_{k+1}) \cdots (x_k - x_n)}$$

We note that there are a total of $n + 1$ of these $L_k(x)$ polynomials, and each contains precisely n factors. Thus, each of them will be a polynomial of degree n. Moreover, each of these $L_k(x)$ is missing the corresponding factors $(x - x_k)$ from the numerator and $(x_k - x_k)$ from the denominator. That is, $(x - x_0)$ is missing from $L_0(x), (x - x_1)$ from $L_1(x)$, and so forth. We now form the expression

$$P_n(x) = y_0 L_0(x) + y_1 L_1(x) + \cdots + y_n L_n(x)$$

Since this is a sum of polynomials of degree n, it is itself a polynomial whose degree is at most n (there might be cancellation of the highest-power terms). Further, $P_n(x)$ passes through each of the $n + 1$ points $(x_0, y_0), (x_1, y_1), \ldots, (x_n y_n)$. To see this, we consider the point $x = x_0$, for example. We see that

$$L_0(x_0) = \frac{(x_0 - x_1)(x_0 - x_1) \cdots (x_0 - x_n)}{(x_0 - x_1)(x_0 - x_1) \cdots (x_0 - x_n)} = 1$$

and, if $k \neq 0$,

$$L_k(x_0) = \frac{(x_0 - x_0)(x_0 - x_1) \cdots (x_0 - x_n)}{(x_k - x_0)(x_k - x_1) \cdots (x_k - x_n)} = 0$$

Using this type of argument, it is easy to see that

$$L_k(x_k) = 1 \quad \text{and} \quad L_k(x_j) = 0$$

for any $j \neq k$. Therefore, for any j,

$$P_n(x_j) = y_0 L_0(x_j) + y_1 L_1(x_j) + \cdots + y_j L_j(x_j) + \cdots + y_n L_n(x_j)$$
$$= 0 + 0 + \cdots + y_j \cdot 1 + \cdots + 0$$
$$= y_j = f(x_j)$$

as required.

We again note that whenever x is at one of the interpolating points $x_0, x_1, \ldots, x_n$, then $P_n(x)$ is in exact agreement with $f(x)$. If x is any other value, then $P_n(x)$ is an approximation to $f(x)$. If x is very close to one of the interpolating points, then $P_n(x)$ will be very close to $f(x)$. On the other hand, if x is far from any of the x_k, then the approximation is usually not as good. That is, the accuracy depends on how close to any of the interpolating points x is.

Furthermore, the accuracy depends strongly on the number of interpolating points used. As a general rule, the more points, the greater the accuracy. Of course, there is a trade-off if too many points are used and the resulting approximating polynomial $P_n(x)$ is of very high degree. While we achieve great accuracy, we also attain a level of possibly unnecessary complication in the resulting calculations using $P_n(x)$, so there may not be a sufficient saving in effort in using $P_n(x)$ to replace a known function $f(x)$.

So far, we have concentrated exclusively on the problem of approximating a given or unknown function $f(x)$ by an interpolating polynomial $P_n(x)$. Once we have $P_n(x)$, however, we can use it for other purposes. For one, if the interpolating polynomial is a good fit to a given curve, then it is reasonable that the area under the original curve should be relatively close to the area under the graph of the interpolating polynomial. That is, the integral of $P_n(x)$ should be a reasonable approximation to the integral of the function $f(x)$,

$$\int_a^b f(x)\,dx \approx \int_a^b P_n(x)\,dx$$

Moreover, it also follows that the derivative of the interpolating polynomial is usually a relatively good approximation to the actual derivative $f'(x)$,

$$f'(x) \approx P_n'(x)$$

As a consequence, in many computational situations involving calculus, we can use the interpolating polynomial in place of the actual function. This was precisely what was done in Section 3.2 when we treated derivatives. Thus, rather than applying the limit process to the difference quotient

$$\frac{f(x + h) - f(x)}{h}$$

that quickly approaches the form $0/0$ as h approaches 0, it was far more accurate to determine an interpolating polynomial of fairly high degree (degree $n = 6$ was used) and calculate the derivative of the polynomial in closed form.

ANSWERS TO
ODD-NUMBERED PROBLEMS

SECTION 2.1

1 0, 1, 2.25, 3.61, 3.9601   **3** 2.449490, 8.124038, 13.26650, 18.330303
5 1, 1.024485, 1.059182, 1.064682

SECTION 2.2

1 on $[-10, 10]$ with $h = .1$: maximum at $x = -10$, decreases to $x = 1.5$, increases to $x = 10$
3 on $[-10, 10]$, $h = .1$: maximum at $x = -10$, decreases from $x = -10$ to $x = -6$, increases to $x = 0$, decreases to $x = .6$, increases to $x = 10$
5 on $[0, 10]$ with $h = .1$: minimum at 0, increases from 0 to 2, decreases to minimum at 10
7 on $[0, 10]$, $h = .1$: minimum at 0, increases to maximum at 10
9 on $[-1, 10]$, $h = .1$: minimum at -1, increases to maximum at 10
11 on $[-10, 10]$, $h = .1$: maximum at -10, decreases to -1, increases to 1, decreases to minimum at 10
13 on $[-10, 10]$, $h = .1$: maximum at -10, decreases to -3, increases to 3, decreases to minimum at 10
15 on $[-3.14159, 3.14159]$, $h = .1$: maximum at -3.14159, decreases to -3.04159, increases to -2.14159, decreases to $-.04159$, increases to maximum at 3.14159
17 on $[0, 10]$, $h = .1$: minimum at 0, decreases to 3.4, increases to 6.4, decreases to 9.5, increases to maximum at 10

SECTION 2.3

1 4, 8, 12, 16, 20, 24   **3** 1/2, 1/4, 1/8, 1/16, 1/32, 1/64   **5** $-9, -2, 17, 54, 115, 206$
7 2/3, 4/9, 8/27, 16/64, 32/125, 64/216   **9** $a_n = 2n + 1$ **11** $a_n = 192/2^{n-1}$, $n = 1, 2, \ldots$
13 $a_n = n/(n + 2)$, $n = 1, 2, \ldots$ **15** 0   **17** no limit   **19** 2   **21** 0   **23** 0   **25** no limit
27 0   **29** no limit

SECTION 2.4

1 7   **3** $-1/24 = -.045454\ldots$ **5** 12   **7** .142857   **9** .666667   **11** .25   **13** .166667
15 1   **17** no limit   **19** 2.718282   **21** 1   **23** 0   **25** 1

SECTION 2.5

1 .4   **3** 0   **5** 0   **7** 1   **9** 1   **11** 1   **13** 0

SECTION 2.6

1 .347168   **3** -1.879028 **5** 2.257080   **7** 1.353210, -1 **9** .860179   **11** 1.165561

SECTION 2.7
---

1 .3, .012, .0006   **3** .0475, .0019   **5** .0109259, .000546296

SECTION 3.1
---

1 0, -3, -108 **3** not defined at $x = 0$, .5, .166667   **5** $-.5$, not defined at $x = -1$ **7** 5, 5
9 0, 1   **11** $y - .761703 = .334254(x - \pi/4)$ **13** does not exist   **15** does not exist

---
SECTION 3.2

See Section 3.1 answers.

SECTION 3.3
---

1 -8, 4492   **3** -10, 122462, 1.68 $E8$ **5** 144, 504, 1104   **7** 1.23913   **9** .991257, $-.727100$
11 $-.363380$

SECTION 3.4
---

11 .6417149   **13** .679194

SECTION 3.5
---

1 $c = -2$ **3** $c = 2$ **5** $c = 3.7641$ **7** $c = 3.75$ **9** $c = 4.1408$

SECTION 3.6
---

1 on $[-20, 20]$: minima near $x = -20$, 1.6667; maxima near $x = -2$, 20
3 on $[0, 10]$: minimum near 0; maximum at 10 (*Note*: Absolute minimum is at $x = -1$, but the computer cannot handle it.)
5 on $[-10, 10]$: minima near -10, -1.99, 1.99, 10; maxima near -2.01, 0, 2.01 (*Note*: Function is not defined at $x = \pm 2$.)
7 on $[-5, 5]$: minima near -2.82, 2.82; maxima near -5, .66, 5
9 on $[0, 5]$: minimum near .065; maxima near 0, 5
11 on $[0, 10]$: minima near 0, 10; maximum near 1.475

SECTION 4.1
---

1 27.84 using $n = 100$ **3** 4.715075 using $n = 100$ **5** .393947 using $n = 100$
7 .342020 using $n = 100$ **9** 2.492981 using $n = 100$ **11** 1.700615 using $n = 100$

SECTION 4.2
---

1 28 using $n = 50$ **3** 4.750300 using $n = 50$ **5** .3926907 using $n = 50$ **7** .337432 using $n = 50$
9 2.493227 using $n = 50$ **11** 1.700574 using $n = 50$

SECTION 4.3

1 28 using $n = 20$ **3** 4.75000001 using $n = 20$ **5** .392699 using $n = 20$ **7** .337365 using $n = 20$
9 2.492907 using $n = 20$ **11** 1.700629 using $n = 20$

SECTION 4.5

1 $c = .6225$ **3** $c = 1.368$ **5** $c = 1.0454$ **7** $c = .7038$ **9** $c = .6702$ **11** $c = 2.3057$
13 $c = .7603$

SECTION 4.6

1 (a) .0865   **(c)** curves intersect at .7083; area = 234.1438   **2 (a)** .9979   **(c)** 6.58   **3 (b)** .9694
4 (a) 1.4794

SECTION 5.1

1 1   **3** .6931472   **5** 2.718282   **7** 1, .5, .333333   **9** .567144   **11** 5.2031

SECTION 5.2

1 1, 403.429, 3. 269017 E6

SECTION 5.3

1	x	y		**3**	x	y
	−.7937	−.5			.0202	1
	0	0			.4428	2
	.7937	.5			1.0737	4
	1	1			1.5033	6
	1.1447	1.5			2.0706	10
	1.2599	2			2.5238	15
	1.7099	5			2.8424	20
	2.4662	15			3.0871	25
	2.7144	20			3.2852	30
	2.9240	25			3.5947	40
	3.1072	30			3.8323	50
	3.4199	40				
	3.6840	50				
	3.9149	60				

SECTION 5.4

1

t	$A(t)$
0	1000.00
1	1056.54
2	1116.28
3	1179.39
4	1246.08
5	1316.53
6	1390.97
7	1469.61
8	1552.71
9	1640.50
10	1733.25
11	1831.25
12	1934.79
13	2044.19
14	2159.77
15	2281.88
16	2410.90
17	2547.21
18	2691.23
19	2843.40
20	3004.17

3 (a) 268.6 feet;   **(b)** 499,551 miles;   **(c)** 22.69 days   **5 (a)** 7.167 hours;   **(b)** 11.98 hours

7

t	$A(t)$
0	1000
1	435.351
2	404.445
3	400.596
4	400.081
5	400.011
6	400.002
7	400
8	400
9	400
10	400

SECTION 5.5

1 (a) 3   **(b)** 3   **(c)** 10   **(d)** -2 **(e)** divergent   **3** maxima near 0, 3.5186; minima near .5341, 2π
5 1.3726

SECTION 5.6

1 (a) 1   **(b)** $-.5$ **(c)** 0
3 curves intersect near (.69287, .74965); $y - .74965 = 1.2498\,(x - .69287)$; $y - .74965 = -.74965\,(x - .69287)$

5 1.7668

SECTION 5.7

1

x	Euler
0.00	20.00
.25	21.25
.5	22.5
.75	23.75
1	25
1.25	26.25
1.5	27.5
1.75	28.75
2	30
2.25	31.25
2.5	32.5
⋮	⋮
8.75	63.75
9	65
9.25	66.25
9.5	67.5
9.75	68.75
10	70.00

3

x	Euler
1.00	4.00
1.16667	5.83333
1.33333	7.97222
1.5	10.4722
1.66667	13.3889
1.83333	16.7778
2	20.6944
2.16667	25.1944
2.33333	30.3333
2.5	36.1667
2.66667	42.75
2.83333	50.1389

5

x	Euler
0.00	6.00
.0666667	7.2
.133333	8.6403
.2	10.3695
.266667	12.4461
.333333	14.9401
.4	17.9355
.466667	21.5333
.533333	25.8545
.6	31.0443
.666667	37.2772
.733333	44.7622
.8	53.7505
.866667	64.5433
.933333	77.502
1	93.0605
1.06667	111.739
1.13333	134.163
1.2	161.081
1.26667	193.393
1.33333	232.179

11

x	Euler	x	Euler
1.0	0.0	3.5	2.95657
1.1	.1	3.6	3.03564
1.2	.218534	3.7	3.11249
1.3	.35183	3.8	3.18725
1.4	.495452	3.9	3.26002
1.5	.645061	4	3.33092
1.6	.796945		
1.7	.948244		
1.8	1.09695		
1.9	1.24175		
2	1.38191		
2.1	1.51705		
2.2	1.64706		
2.3	1.77201		
2.4	1.89206		
2.5	2.00742		
2.6	2.11835		
2.7	2.22509		
2.8	2.32791		
2.9	2.42705		
3	2.52274		
3.1	2.61519		
3.2	2.70463		
3.3	2.79122		
3.4	2.87515		

SECTION 6.1

1 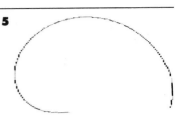   **3**   **5**

7 $Q = 0$, $R = 0$ and $Q = .6435$, $R = .6435$
9 $Q = 3.27$, $R = 1.016$ and $Q = 4.39$, $R = 2.375$ (*Note*: There are two other points where the curves cross, but not for the same values of the angle Q.)

SECTION 6.2

1 1.7548   **3** 6.2832   **5** .7498
7 Intersection at $Q = 1.199$, $Q = 2.862$. Total area $= 4.4584 + .1472 + 3.4421 = 8.0477$ **9** 22.6274

SECTION 6.3

1 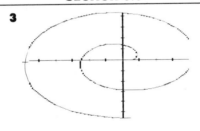   **3**   **5**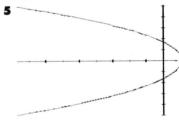

7 80.00

SECTION 7.2

1 convergent   **3** convergent   **5** convergent   **7** divergent   **9** divergent

SECTION 8.1

1 1, 1, .5, .166667, .041667, .008333, .001388, .000198, .0000248, .00000276
3 1, 2, 3, .333333, .333333, .3, .022222, .015873, .010714, .000617
5 1, 3, 13, 59, 269, 1227, 5597, 25531, 116461, 531243

SECTION 8.2

1 convergent   **3** divergent   **5** divergent   **7** convergent   **9** convergent   **11** convergent
13 divergent   **15** convergent   **17** 7.389056, 54.59815, .0067379   **19** $-.111111$, 1

SECTION 8.3

1 .7193398 using $n = 3$ **3** 1.0304545 using $n = 3$ **5** .01745239 using $n = 5$
7 .40729167 using $n = 5$

9

x	$P_5(x)$
0	0
.0785	.1293
.1571	.2596
.2356	.3919
.3141	.5274
.3927	.6672
.4712	.8129
.5498	.9662
.6283	1.1289
.7069	1.3036

11

x	$P_5(x)$
.5236	.7146
.6283	2.0565
.7330	3.3983
.8376	4.7402
.9425	6.0821
1.0472	7.4240

13 Taylor: .9461 using $n = 5$; Simpson: .9460 on $[.0001, 1]$
15 Taylor: .303 using $n = 4$; Simpson: .2970 on $[.0001, .3]$

SECTION 9.1

1 $-2/3$ **3** no limit   **5** no limit   **7** 2   **9** 0   **11** 0   **13** no limit   **15** no limit
17 no limit

SECTION 9.2

1 $-7, 5128$ **3** $e, e, -1/e$ **5** minimum at $(1, -1)$; saddle at $(0, 0)$
7 minimum at $(-2, \sqrt{3})$; saddle at $(-2, -\sqrt{3})$ **9** minimum at $(2, 1)$
11 minimum near $(1.48187, 1.85949)$; saddle at $(0, 0)$ **13** maximum at $(-2, 3/2)$
15 maxima at $(1, \pm\pi/2)$, $(1, \pm5\pi/2)$; minima at $(1, \pm3\pi/2)$; saddle points at $(0, 0)$, $(0, \pm\pi)$, $(0, \pm2\pi)$, $(0, \pm3\pi)$

SECTION 9.3

1 .218713 using $n = m = 100$ **3** 1.12500 using $n = m = 100$
5 3.96735 using $n = m = 100$ **7** .98209 using $n = m = 100$
9 .795872 using $n = m = 100$ **11** 38.8976 using $n = m = 100$

SECTION 10.1

1   **3**   **5**

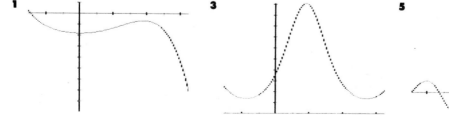
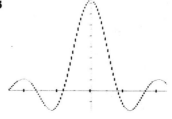

7

9

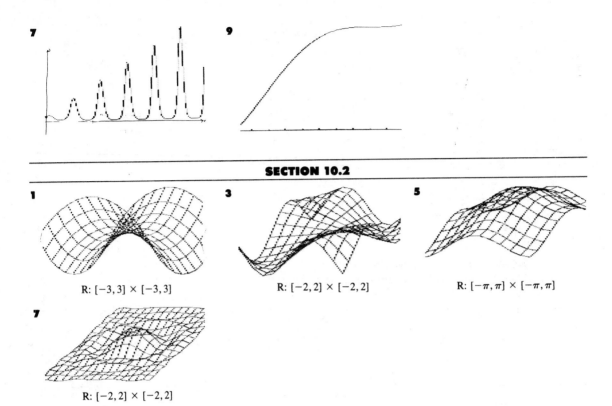

SECTION 10.2

1

R: $[-3,3] \times [-3,3]$

3

R: $[-2,2] \times [-2,2]$

5

R: $[-\pi,\pi] \times [-\pi,\pi]$

7

R: $[-2,2] \times [-2,2]$

INDEX